# Swamp Rat

AMERICA'S
THIRD
COAST

Carl A. Brasseaux and Donald W. Davis, series editors

# SWAMP RAT

The Story of Dixie's Nutria Invasion

Theodore G. Manno    Foreword by Elaine Miller Bond

University Press of Mississippi / Jackson

This contribution has been supported with funding provided by the Louisiana
Sea Grant College Program (LSG) under NOAA Award #NA14OAR4170099.
Additional support is from the Louisiana Sea Grant Foundation. The funding
support of LSG and NOAA is gratefully acknowledged, along with the matching
support by LSU. Logo created by Louisiana Sea Grant College Program.

www.upress.state.ms.us

The University Press of Mississippi is a member
of the Association of American University Presses.

First printing 2017

∞

Library of Congress Cataloging-in-Publication Data

Names: Manno, Theodore G., author.
Title: Swamp rat : the story of Dixie's nutria invasion / Theodore G. Manno ;
foreword by Elaine Miller Bond.
Description: Jackson : University Press of Mississippi, 2017. | Includes
bibliographical references and index. | Description based on print version
record and CIP data provided by publisher; resource not viewed.
Identifiers: LCCN 2016051251 (print) | LCCN 2017011425 (ebook) |
ISBN 9781496811950 (epub single) | ISBN 9781496811967 (epub institutional) |
ISBN 9781496811974 ( pdf single) | ISBN 9781496811981 (pdf institutional) |
ISBN 9781496811943 (hardcover : alk. paper)
Subjects: LCSH: Coypu—Gulf States. | Nonindigenous pests—Gulf States. |
Wetland ecology—Gulf States.
Classification: LCC QL737.R668 (ebook) | LCC QL737.R668 M36 2017 (print) |
DDC 577.68—dc23
LC record available at https://lccn.loc.gov/2016051251

British Library Cataloging-in-Publication Data available

For the people of America's Third Coast,
who remain resilient in the face of a changing environment.
AND
For the millions of nutria that were exterminated
after their ancestors fought bravely to survive in an unfamiliar land;
they acted as Nature made them, knowing nothing of our anger.
What happened was our fault, not theirs.

Up Silver Street he went, and down Gold Street, and at the end of Gold Street is the harbor and the [strait] beyond. And as he paced along, slowly and gravely, the townsfolk flocked to door and window, and many a blessing they called down upon his head.

As for getting near him there were too many rats. And now that he was at the water's edge he stepped into a boat, and not a rat, as he shoved off into deep water, piping shrilly all the while, but followed him, splashing, paddling, and wagging their tails with delight. On and on he played and played until the tide went down, and each master rat sank deeper and deeper in the slimy ooze of the harbor, until every mother's son of them was dead and smothered.

—Translated from *The Pied Piper of Hamlin, as told by the Brothers Grimm*

# Contents

# Foreword

Theodore ("Theo") Manno and I met in 2005 in the backseat of a car. I know how that sounds. But we were covered in layers of polar fleece, along with hats and scarves that nearly obscured our faces. We were loaded with armfuls of field gear: he carried a walkie-talkie, binoculars, datasheets, and even more articles of warm clothing; I held a camera with a long zoom lens. As for the automobile, it had a picture of a prairie dog laminated to the outside of its doors and a wooden sign posted across the back window: "Prairie Dog Squad Car." We were on a mission of behavioral ecology.

Manno and I and a small "squad" of research assistants spent the winter-spring field season in the mountains of southwestern Utah, at a study site operated by Dr. John L. Hoogland. (Manno, in fact, worked there for two seasons.) Our goals were to assist Dr. Hoogland in documenting the fascinating behaviors of the Utah prairie dog (*Cynomys parvidens*), a species that is threatened with extinction.

Our experience grew into a book, *The Utah Prairie Dog: Life among the Red Rocks* (University of Utah Press, 2014), featuring a foreword by Dr. Hoogland, photos taken by me, and artful writing by Manno. I've always respected Manno for taking on the challenge of authoring a book about prairie dogs, controversial digging mammals that, on the one hand, are beloved for their barking, kissing, and chipper tail-wagging, and, on the other hand, have been killed as "pests." Indeed, prairie dogs are in dire need of understanding, before it's too late.

Now Manno has turned his mind and his pen to nutria, another mammal encumbered by controversy. Just like his first book, *Swamp Rat* is artfully written, engaging, balanced, sometimes funny, other times poetic, always impeccably researched. (The bibliography alone includes hundreds of entries!) It's also rodent-centered, which, in my mind, is a very good thing.

Yet the rodent at the core of Manno's second book is not a rare native keystone species critical for the health of the prairie ecosystem, as is the prairie dog. The nutria, or "swamp rat," is an invasive exotic species in North America, introduced by humans to sensitive habitats where, through no fault of its own, it is inflicting serious environmental and economic damage. In other words, this book is not meant to be "For the Love of Swamp Rats." It's also not "What Do We Do about All These Damn Swamp Rats?!" It's a lesson far more complex.

Stories within the story of *Swamp Rat* track the rise and (thankfully, in my opinion) the decline of the American fur trade; the near-eradication of beavers, muskrats, alligators, and other native animals along the way; the introduction, or "unnatural history," of nutria in Louisiana's fragile wetlands and beyond; and the present-day efforts to keep these South American rodents, so far from their home, from, well, "eating up the place."

Some of these efforts, especially hunting and trapping and bounty programs, are difficult to accept. It pains me, as a vegetarian for more than 26 years, to learn of any animal suffering, especially as a result of humanity's own mistakes.

But, ultimately, I trust Manno. I trust that he's uncovered all current knowledge about the problem of invasive nutria, that he's considered it very carefully, and that he's balanced his scientific training with his animal-loving nature. Readers, I'm sure, will trust him too, once they experience the power and intricacy of his writing.

And so I'm holding great hope for *Swamp Rat*—that it will serve as a resource for prevention. Perhaps it will also inspire scientists and others to seek new, humane solutions to this and other environmental issues. And maybe, just maybe, it will move some readers, like it moved me, toward reevaluating our relationship with animals. Indeed, one might draw many lessons from this book about a soft, furry creature with orange teeth and a ratlike tail—a story that twists and turns and, like the waters of the Louisiana bayou itself, runs muddy and chest deep.

—Elaine Miller Bond
Author/Photographer
November 29, 2015

# Preface and Acknowledgments

I have been interested in rodents for as long as I can remember. My curiosity began with the gray squirrels (*Sciurus carolinensis*) in the backyard of my childhood home in New Jersey, scurrying here and there to bury their acorns, sometimes leaping from one tree to another. My own style of leaping was more metaphorical, as I hopped between majors during my first two years of collegiate study. But my interest in rodents eventually motivated me to acquire a degree in biology when other endeavors did not. For example, self-proclamation of my academic major as "pre-vet," or some such, was stymied quickly when the realities of working at a veterinarian's practice kicked in.

After college, I progressed to the only feasible option for a biology graduate who is uninterested in clinical studies or allied health—graduate school—during which I conducted research on the behavior of various ground squirrels (*Cynomys* and *Urocitellus* spp.), including prairie dogs in Utah, which were once on the brink of extinction. Unlike my short-lived foray into veterinary studies, field research on prairie dogs was an interest that rendered me strange in the eyes of locals. This was because Utah prairie dogs (*Cynomys parvidens*) were long-reviled beasts; they were systematically annihilated during western settlement in the 1800s and 1900s because of their grass-feeding proclivities. Most folks assumed that prairie dogs competed with livestock for forage, and ranchers believed that cattle (*Bos taurus*) and horses (*Equus callabus*) broke their legs in prairie dog burrows.

But despite the popular opinion that they were pests that impinged on ranchland, Utah prairie dogs were (and still are) protected by the Endangered Species Act (ESA) and could not be killed or relocated without strict governmental oversight. This conflict between citizens and wildlife officials, which drummed up arguments about land stewardship and property own-ership rights, did not seem to faze my rodent brethren in the least. They continued munching on Utah's short grasses merrily, and failed to under-stand the charges against them or the anger they caused by not adhering to our demands.

The same failure to understand human constructs applies to Mother Nature, and the nation learned this in 2005 when Hurricane Katrina slammed into the Gulf Coast and left death and destruction in her wake. I was conducting PhD research at Auburn University in Alabama when Katrina made landfall and I was thankful to be around 300 miles northeast of the storm's eye, but living in an affected state heightened my interest in Katrina's aftermath and the swell in Atlantic tropical cyclone activity that followed. As the worst hurricane season in American history unfolded, I read all that I could about a bizarre news story. A few scientists were studying whether large rodents called *nutria* (*Myocastor coypus*) were compounding human-induced coastal erosion along the Gulf Coast by devouring grasses that hold soil together. The concern was that nutria feeding was increasing shoreline exposure to the storm surges from hurricanes.

The reports mentioned that nutria were brought to the United States from South America to stimulate the commercial fur trade, and they began devouring vegetation after being released or escaping from enclosures. Nutria were still eating coastal flora when, a few years after Katrina, addi-tional hurricanes battered the Gulf Coast region and storm surges caused thousands of nutria corpses to wash up on beaches. Considered by most area residents to be a token bright spot in an otherwise disastrous series of storms, the deaths of the destructive nutria were sometimes celebrated.

As a burgeoning rodentologist, it was hard for me to ignore a story that featured a large, ravenous rodent. The tale was also an interesting departure from my previous wildlife-related experience. Prairie ecosystems are disap-pearing because, among other issues, the prairie dogs on which dozens of carnivores depend for prey now inhabit just 5 percent of their former range. The nutria situation posed an entirely different issue, as too many nutria existed in a region that was never intended to be their residence—even after thousands of dead nutria washed up on America's beaches. So, as the years came to pass, I monitored environmental goings-on involving nutria and my interests continued to evolve. Eventually, what began as a casual investiga-tion into the effects of nutria feeding behavior on the environment led me to study analyses of the commercial fur trade, commentaries on modern wildlife management philosophy, debates about coastal economic issues, and simulations of nutria feeding interactions with hurricanes, floods, and levees.

I encountered many excellent sources of information on nutria history and biology, but I did not find a widely accessible book-length source that provided a definitive and comprehensive report of nutria-related knowledge. Responding to this gap in the literature, I conceived of a book about nutria four years ago. My rationale for writing the book was for folks to understand how our coastal ecosystems suffered from the introduction of exotic nutria so that we could stop history from repeating. But the task seemed too herculean after I realized that an analysis of the international fur trade during the past three centuries was necessary to tell the story of Dixie's nutria invasion. So, I originally dismissed the idea as one that would never come to fruition because of other commitments.

As unsustainable housing development, additional hurricanes, and the worst oil spill in history continued to batter the Gulf Coast, shoreline health transformed from a regional issue into a national controversy. It eventually became clear that I could no longer ignore the absence of a book devoted completely to nutria and their effects on the wetland ecosystems to which they were introduced. Thus, I decided that my goals for *Swamp Rat* would be simple, but ambitious—to write the definitive source on nutria, and to make it readable for amateur naturalists and professional biologists alike. I wanted a book that could sit on office bookshelves or a nightstand, and one that represented not just an academic effort, but also a popular account that used substantial scholarly findings to express the bizarre story of nutria as an invasive species. And I wanted the book to be enhanced by rich images that would be interesting for a general audience.

For interested laypeople, reliable and unbiased information from professional biologists can be difficult to obtain. When found, the material is often presented with specialized or technical language. Thus, my approach is to tell the story of Dixie's nutria invasion as an independent writer and researcher with a scientific background, rather than as a biologist who is writing about his direct involvement with an environmental issue. I have therefore made every effort possible to seek input from folks who are directly involved with nutria-related environmental management and initiatives, such as officials involved with coastal restoration and concerned citizens who are fourth- or fifth-generation fur trappers. And I have read every available primary research document involving nutria, including records from more than 80 years ago pertaining to their introduction.

For several years, I have been almost completely ensconced in achieving my goal of a comprehensive book about nutria. Besides requiring years to research and write, *Swamp Rat* has taken me across the Deep South, from the Mississippi shoreline to the forests of Cameron Parish and the ravaged coast of New Orleans; through western Europe and the parched desert of southern Arizona; and into zoological facilities, libraries, and archived collections where I indefatigably researched nutria-related documents. This book on swamp rats has kept me, well—*swamped*, pun intended. It has been a long road, but the people I have come to know as a result of the writing process have made this endeavor worthwhile. Many individuals have offered their assistance, their time, and, most important, their inspiration.

Acquiring editor Craig Gill, series editors Carl Brasseaux and Don Davis, and the entire editorial team at the University Press of Mississippi recognized the importance of this project. They provided outstanding assistance with preparation and production of the manuscript, and I am grateful to all of them.

Copy editor Anne Rogers spared my readers from potentially embarrassing typographical errors, inconsistencies, and other imperfections in my writing. I thank her for a thorough review that greatly improved the quality of this book.

Two peer reviewers graciously accepted the considerable responsibility of providing constructive criticism about the manuscript. Their astute comments helped me improve the book, and I thank them for their willingness, availability, and professionalism.

State officials and participants in nutria-related initiatives were helpful to me while writing *Swamp Rat*, and I appreciate their assistance immensely. Jennifer Hogue and Edmond Mouton from the Louisiana Division of Wildlife and Fisheries (LDWF) provided information on the Coastwide Nutria Control Program (CNCP). Cree McCree, founder and project director of Righteous Fur, talked to me about nontraditional uses for trapped nutria from the CNCP, such as clothing and food. Jacoby Carter, research ecologist for the United States Geological Survey (USGS) at the National Wetlands Research Center (NWRC) in Lafayette, Louisiana, was a valuable resource during the course of my research. And Chef Philippe Parola gave an extremely compelling interview regarding his ideas for human consumption of nutria meat on an unforgettable Friday afternoon in Baton Rouge.

Shane K. Bernard, historian and curator for McIlhenny Company, guided my trip to the Tabasco archives on Avery Island and helped me improve the third chapter with his sweeping knowledge of the area's history and nutria introductions in Louisiana. I cannot thank the McIlhenny family enough for allowing me to interview Dr. Bernard and for having me as a guest at the Tabasco facilities and Jungle Gardens.

For assistance with locating images and for granting permission to publish materials from library collections, I thank Judy Bolton from Special

Collections at the Louisiana State University (LSU) libraries; Charlene Bonnette, preservation librarian for the State Library of Louisiana; and I. Bruce Turner and Jane Vidrine from the University of Louisiana at Lafayette's Edith Garland Dupré Library. For directing me to documents pertaining to the nutria trade in the H. Conrad Brote collection, I thank Lindsey Reno, reference librarian for the Earl K. Long Library at the University of New Orleans. I also thank Shane K. Bernard, the McIlhenny family, Cree McCree, Jennifer Hogue, the LDWF, the CNCP, and Chef Philippe Parola for the use of photographs, graphs, tables, recipes, and other materials.

For general support throughout my writing career, I thank photographer, artist, and writer Elaine Bond, who wrote the foreword for this book and did a marvelous job with documenting prairie dog behaviors for my first book. And for miscellaneous assistance, I thank my Siamese cat, Mochi, who also loves rodents, but probably not for the same reasons that I do.

Most of all, I thank the people of America's Third Coast for their support during this project, and for inspiring us all by pressing on through recent events that have brought some tough times. May they endure.

—T. G. M.
February 20, 2016

# Swamp Rat

Chapter 1

# An Unnatural History

The coypu is a native of Chil[e]. . . . It is one of the aquatic animals, though it does not live so constantly in the water as the beaver. . . . It is of moderate size, averaging two feet in length from the nose to the root of the tail, while the tail measures rather more than a foot. Its colour is rich, shining brown . . . due to the long hairs which penetrate through the grayer wool with which the body is thickly covered, and which lie thickly upon each other like the straws in a thatched roof. . . . This is one of the animals which have tended to decrease the demand for beaver skins, its fur being exclusively used in the manufacture of hats. . . . I believe that a French furrier, M. Becheur, was the first who discovered the use that could be made of the fur, and by him alone as many as twenty thousand skins have been imported in a single year. Before that time, very little was known of the animal.

—**Rev. J. G. Wood,** *Beeton's Brave Tales, Bold Ballads, and Travels and Perils by Land and Sea* (1872)

*On a warm, humid morning in early September 2012, US Route 90 stretches across the southern Mississippi coastline, connecting historic towns with green bayous. Just 50 miles west is the birthplace of American jazz and the excitement of Bourbon Street, the French Quarter, and Café du Monde—a city like no other. In its heyday, New Orleans was the most populous city in the South and an inspiration to twentieth-century writer Tennessee Williams, who wrote about the "desire" of its people from a dilapidated upstairs apartment on Saint Peter Street. And US-90 was one of the most scenic thoroughfares in the country, offering views of the Gulf of Mexico to the south, antebellum mansions to the north, and tall, elegant oak trees throughout.*

*But roads through this region now feature deserted homes, destroyed build-ings, under-construction replacement bridges, and fallen vegetation as resi-dents struggle in the face of several hurricanes and a major offshore oil spill. The summer of 2012 has brought another storm named Isaac to a region still not recovered from $108 billion worth of damage in 2005 following Hurri-cane Katrina's swirling winds and torrential rains. Even as a category 1 storm, Isaac has caused mandatory evacuations of all locations in Mississippi south of US-90 and 1,500 troops from the National Guard have been deployed to the state's three southernmost counties.*

*Property damage has not been the only issue brought on by the recent hur-ricanes. Massive tidal surges from cyclonic winds bring drowned animals onto the beach to burst open and rot away, creating awful sanitation problems. Mississippi saw the problem before with Hurricanes Katrina and Gustav, but Isaac's tides have beached over 16,000 corpses of an odd-looking mammal with orange teeth, the most from a storm yet. The coast smells from their decaying flesh, and officials with specialized training in hazardous-waste disposal have been called in for duty.*

*Under normal circumstances, this number of animals washing ashore after a storm would be considered tragic by almost everyone. Seals, dolphins, whales, or even fish and sea stars would be mourned by the environmentally conscious and even some folks who are usually indifferent to wildlife. Instead, the prevail-ing attitude is "good riddance," and the death of these beasts is widely regarded as a token bright spot of an active hurricane season. Some even blame the now-deceased animals for exacerbating damage from recent storms by eating away coastal wetlands—land ecosystems that are permanently or seasonally satu-rated with water and serve as a buffer between storm surges and residential areas. In an otherwise frustrating situation, almost no one is upset about the drownings—except maybe the workers charged with removing the dead bodies from the beach.*

*The reason for this is clear. These are not the typical sea mammals that folks know and love. These are members of America's most notorious invasive spe-cies—nutria, callously referred to by some locals as* swamp rats.

✦ ✦ ✦

At a first glance, nutria seem like fascinating, charming, and likeable ani-mals. Peacefully swimming in their watery home and mundanely foraging on plant life, their presence hardly seems capable of historical significance. With an unusual, ratlike appearance that features a chubby, pear-shaped body and webbed toes, nutria can baffle unsuspecting visitors to wetlands

All in the family. Nutria reproduce quickly, and they have proliferated wildly across the southeastern and northwestern United States. Courtesy of Shutterstock, image #216921022.

who see them swimming through the swamp. Nutria also seem hardworking and resourceful as they maintain their burrows and nests near the water's edge, and their social interactions, complete with a wide repertoire of vocal and scent-based communications, can make them endearing to amateur naturalists.

But the benignity of nutria is gone forever, as they are now known widely for two behaviors—voracious appetite for plant life and reproductive prowess—that guarantee a bad reputation. Nutria became famous for these distinctions after the popularity of beaver-felt hats brought early European fur traders to North America in the seventeenth century. Fierce battles ensued between economic powers like England, France, and the Netherlands to lay claim to untapped fur resources, resulting in an active American fur trade that collapsed in the 1830s because of overtrapping. Soon afterward, nutria were imported from their natural range in South America to the United States as a viable replacement that was cheaper and more plentiful.[1]

In the late 1930s, nutria were introduced to Louisiana's wetlands when they escaped or were released from captive breeding facilities. This dramatically changed the region's environment. Nutria multiplied quickly and expanded their geographic range while threatening to displace the native muskrats that were the lifeblood of the area's fur industry. State officials responded by facilitating a market transition away from muskrat and transplanting or protecting the nutria. Thus, nutria pelts became a major component of Louisiana's economy, especially during the 1970s when European companies offered luxury products made with nutria fur. But once fur fashions were no longer *en vogue*, the nutria in Louisiana's wetlands multiplied, unharvested and unabated.[2]

Prolific reproduction continued until nutria earned the title of *invasive species*, which refers to an organism that is nonnative and adversely affects the habitat to which it is introduced.[3] Forming burrow-filled havens along stretches of water, nutria munched plant life and wasted 90 percent of their forage as they chewed only on succulent roots. When anti-fur sentiment reached an all-time high in the 1980s, lower nutria harvest levels brought complaints of chronic damage to wetlands from land managers.[4] One estimate stated that Louisiana's nutria population increased from 20 individuals to 20 million in only two decades and was contributing to the loss of a football field's worth of wetlands daily.[5]

This story repeated itself wherever fur farmers released their nutria stock. In Maryland, the Blackwater Unit of the Chesapeake Marshlands National Wildlife Refuge (CMNWR) experienced a growth in the nutria population of over 300 percent during 1968–98, and a 50 percent loss of wetlands resulted. Nutria in the state of Washington outcompeted native muskrats for food and habitat while destroying root "mats" of vegetation that held wetlands together, leading to erosion from wind and waves. And nutria in Oregon robbed vegetable gardens, destroyed lawns, swam in restored ponds, fought with family pets, and nibbled water pipes.[6] But despite the damage that nutria caused nationwide, ground zero for the destruction was the Mississippi delta, where locals dubbed nutria as "the rat that ate Louisiana."[7]

Today, wildlife managers continue their struggle to restore nutria-damaged wetlands—especially in coastal Louisiana, where nutria have contributed to the loss of a wetland area roughly the size of Delaware.[8] Progress has come from massive control programs, most notably a bounty in Louisiana that introduces harvest incentives by paying hunters five dollars per nutria tail.[9] And hurricanes have also taken their toll on nutria, washing tens of thousands of their corpses onto Gulf Coast beaches. But nutria remain persistent and continue to cause mayhem worldwide, taking their place alongside brown tree snakes (*Boiga irregularis*), which annihilated native bird species in Guam; feral pigs (*Sus scrofa*), which eat bird eggs and trample native plants in Hawaii; and other notorious invasive animals.[10]

Close-up. A close-up of a nutria from the slide collection of the Louisiana Department of Wildlife and Fisheries. Courtesy of Louisiana State University Special Collections: Louisiana Department of Wildlife and Fisheries Slide Collection.

The story of nutria introduction to the United States is fascinating, bizarre, and tragic. It is a saga that touches four centuries of the Gulf Coast's economic and cultural history, entrenched in the overarching themes of progress, individualism, conquest, and fortune. From the involvement of nutria in several revitalizations of the fur industry, to their interference with Louisiana's valuable sugar production, to their potential ecological interactions with Hurricane Katrina, to their bizarre connection with a well-known spicy condiment called Tabasco sauce, the history of nutria is intertwined tightly with American history. But the story of nutria in North America is rife with misconceptions regarding fundamental topics such as what nutria eat, the type of environmental damage they inflict, and how they entered the wild and proliferated. For the past few years I have been engrossed in researching every detail of what I call the "nutria invasion," with the goal of completely understanding how nutria were introduced to American wetlands and what happened afterward. To achieve this goal, I have observed nutria in the field, interviewed wetland restoration specialists and historians, and read every article on nutria that I can find.

Among many questions, I wanted to know why apex predators like alligators were unable to contain the growing population of introduced nutria, and how nutria replace themselves so quickly. I also wanted to find out why Louisiana became a hub of the international fur industry and eventually suffered most from the introduction of nutria; whether twentieth-century anti-fur campaigns directly led to the release of nutria from fur farms; and what environmental officials are doing to restore the wetlands consumed by nutria. I was very intrigued as to why wildlife managers would encourage the transport of nutria into areas where they had few natural enemies, apparently with minimal assessment of the potential consequences. And I was totally captivated by stories of nutria making their way into the wilds of the southern United States. Were the original nutria that led to invasive populations purposely released, or did they escape from their enclosures? If they were released, which fur farmers participated? Are legends about the irresponsibility of these nutria owners true, or do some of these folks take undue blame for unleashing the nutria "plague" on humankind?

While I hope to communicate that nutria, although widely disliked, are fascinating animals with interesting behaviors, I also write about the environmental and economic problems that their introductions have caused. Focusing on Louisiana, and touching on other American and international wetlands, I describe how the eating style of nutria makes them so destructive, and talk about programs that reduce nutria populations or address nutria-related damage. Another theme is how the conquest of North America by Europeans led to the introduction of nutria. I describe how beavers and muskrats brought early European fur traders to the American colonies and territories, and how the Louisiana Purchase expanded trapping opportunities and made the southern Mississippi River a hub for import and export. And then I describe how a collapse of the beaver market in the 1830s eventually led to an influx of captive nutria in the United States to satisfy the strong demand for fur.

Finally, I write along the theme of "never again." I point out that Manifest Destiny created the greediness that compelled settlers to hunt the furbearing beaver to near extinction and ultimately led to furriers hedging their bets with imported nutria. But I also note the irony of how the modern anti-fur movement has unintentionally exacerbated the spread of nutria by leaving their populations poorly harvested. I analyze various solutions to save our wetlands from hungry nutria such as exclusion barriers, statewide bounty programs, and using nutria for human consumption. And then I emphasize that poorly conceived nutria introductions and other mismanagement ultimately resulted in the disappearance of large areas of wetlands that will likely never return.

This is the story of one of the greatest environmental disasters in history. It is the story of a delicate ecosystem that is crucial for shoreline stability and may never be the same again. And although they are considered the villains, it is also the story of millions of nutria that were slaughtered not because they acted wrongly, but because humankind made the mistake of turning their ancestors loose in an area that Mother Nature did not intend to be their home.

The cumulative effect of misguided and environmentally unsustainable human activity continues to negatively impact American wetlands. Restoration and nutria-control programs have been effective and continue to show promise, but the compromised health of wetland ecosystems persists as an issue of national concern. As exotic nutria continue to eat away at Louisiana, I hope that this book highlights the mistakes that caused this debacle so that they are never repeated.

✦ ✦ ✦

Nutria are not rats. They are large, herbivorous, gregarious, and semiaquatic rodents (order Rodentia) that constitute the only member of family Myocastoridae. Like all rodents, nutria (*Myocastor coypus*) have furry bodies and two large front teeth that are constantly growing and in need of sharpening, which is accomplished by gnawing on organic material.[11] Fossil records indicate that nutria have a common ancestor with spiny rats (family Echimyidae) in the Neogene going back to about 7.3 million years ago, and they are closely related to other South American rodents of parvorder Caviomorpha (infraorder Hystricognathi) like capybara, chinchillas, viscachas, tuco-tucos, agoutis, and guinea pigs.[12] Nutria and their ancestors share some characteristics with rats such as large eyes, scaled tails, pointed snouts, and prominent whiskers, but they are not immediate descendants of rats. Although the term *rat* is not taxonomically specific, scientists generally regard only members of the genus *Rattus* to be "true" rats, and the larger body size and broader skulls with larger and higher-crowned teeth of nutria differentiate them from their rat cousins.[13]

Because nutria share some physiognomies with rats and other small mammals, folks often confuse them with certain furbearers—animals with fur that is valued commercially—such as beavers, muskrats, otters, and minks. Inset 1.1 details the differences between these animals so that amateur naturalists can tell if the mammal they are viewing is actually a nutria.

Nutria are portly and pear-shaped, measuring about 1.5–2 feet (45–60 centimeters) long (excluding the tail). Their large, triangular head features a white-patched, tapering muzzle with 4-inch-long whiskers, and they have highly arched backs, short legs, and long, round, scantily haired tails that are around 13–16 inches (35–40 centimeters) long. The pelage of nutria consists

## COMPARISON OF NUTRIA TO OTHER FURBEARING MAMMALS

| | Nutria (Coypu) (*Myocastor coypus*) | North American beaver (*Castor canadensis*) | Muskrat (*Ondatra zibethicus*) | River otter (*Lutra canadensis*) | American mink (*Neovison vison*, formerly *Mustela vison*) |
|---|---|---|---|---|---|
| **Range** | Central Bolivia and southern Brazil to Tierra del Fuego (introduced to Europe, Asia, Africa, and North America) | All of North America except extreme northern Canada, peninsular Florida, and deserts in the Southwest USA and Mexico (introduced to Eurasia) | Almost all of North America to the Gulf Coast and the Mexican border (introduced to Eurasia and South America) | Before settlement, almost all of present-day USA and Canada; now absent or rare in the Southwest and most areas of the prairie states | Throughout the USA except the desert Southwest; throughout Canada except the Arctic coast and some offshore islands (introduced to Europe and South America) |
| **Altitude** | Usually under 1,000 feet (305 meters), but sometimes up to 3,900 feet (1,190 meters) | Under 1,000 feet (305 meters) | Under 1,000 feet (305 meters) | Under 1,000 feet (305 meters) | Usually under 1,000 feet (305 meters) |
| **Guard Hairs** | Yellow-brown to reddish-brown | Reddish-brown or blackish-brown | Dark-brown, sometimes reddish-brown | Light-brown to black | Dark-brown with white patches on the chin, chest, and throat |
| **Underfur** | Brown or dark gray | Lead gray | Dark-brown, sometimes reddish-brown | Light-brown to black | Dark-brown with white patches on the chin, chest, and throat |
| **Size (length does not include tail)** | 1.5–2 feet (45–60 cm) 13–15 lbs (6–7 kg) | 3.2–3.9 feet (100–120 cm) 25–35 lbs (11–26 kg) | 1.5–1.8 feet (45–55 cm) 1.5–4 lbs (0.7–1.8 kg) | 2–3.5 feet (66–107 cm) 11–31 lbs (5–14 kg) | 1.6–2.25 feet (49–68 cm) 1.5–3.5 kg (0.7–1.6 kg) |
| **Tail** | 10 inches (25 cm), long, round, and scaly, sparsely covered in hairs | 12 inches (30 cm), broad, oval-shaped, flat, and scaly with no hairs | 9 inches (23 cm), thin, ribbonlike, and scaly with fine black hairs | 12–20 inches (30–51 cm), very long, thick, and tapered with fine, dense fur | 6–8 inches (15–20 cm), long and bushy |
| **Tracks** | Forefeet: four unwebbed digits with vestigial thumb Hindfeet: five digits, connected by webbing except fifth | Forefeet: five unwebbed digits with distinct claws Hindfeet: five digits, connected by webbing except second is sometimes split | Forefeet: four unwebbed digits with distinct claws Hindfeet: five digits, partially webbed with distinct claws and hairs along margin of toes | Forefeet and hindfeet: five webbed digits with distinct claws | Forefeet and hindfeet: five un-webbed digits |
| **Droppings** | Cylindrical, 2 x 0.5 inches (5 x 1.25 cm) with parallel grooves | Sawdust-like and in large, loose piles | Bean-shaped, around 1–2 x 0.5 inches (2–5 x 1.25 cm), usually deposited in small piles | Loose and tar-like | Greenish-black, with unusually strong odor; cylindrical and long (around 0.3 inches or 0.8 cm) |
| **Den** | Burrows with beds of cut vegetation | Large lodges made from branches and mud | Huts with mounds of vegetation and mud | Holts (dens) constructed in the burrows of other animals or from natural hollows in river banks or under logs | Long burrows in riverbanks, holes, tree stumps, or hollow trees; sometimes formed in crevices, drains, and nooks under stone piles, bridges, or logs |

| | | | | | |
|---|---|---|---|---|---|
| **Social and breeding system** | Gregarious with polygynous family groups; polyestrous | Family groups with monogamous pairs and their offspring; mates once during winter | Family groups with monogamous pairs and offspring; polyestrous | Transient males and variable family groups that may include a female and her offspring, or unrelated individuals of any age; adult males sometimes form social groups | Transient individuals; promiscuous mating occurs once during winter |
| **Usual age of sexual maturity** | Around 6 months | Around 1.5 years | 5–12 months | Males: 2 years Females: 1–2 years | 10 months |
| **Gestation** | 126–141 days | Around 3 months | Usually 28–30 days, sometimes 19–27 days | 61–63 days | Usually around 51 days; range is 40–75 days |
| **Lactation** | 50–55 days | Around 6–8 weeks | Around 28 days | Around 3 months | 5–6 weeks |
| **Number of offspring per litter** | Usually 4–6, can range from 1–13 | Usually 3–4 | Usually 6–7, can range from 3–10 | Usually 1–3, can range from 1–5 | Usually 4, can range from 1–8 |
| **Diet** | Wide variety of aquatic vegetation | Tree bark and some water vegetation | Wide variety of aquatic vegetation, occasionally small fish and reptiles | Fish, occasionally small birds, insects, frogs, and crustaceans | Fish, frogs, shrews, rabbits, earthworms, snails, insects, mice, muskrats, baby ducks, and other waterfowl |
| **Vocalizations** | Piglike grunts, shrill cries | Low groaning sounds | Squeaks and squeals | Snarling growls; hissing barks; shrill whistles; low, purring grunts; snorts; birdlike chirps | Hissing, screams, barks, purrs, chuckling sounds |
| **Type of Water** | Usually fresh, sometimes brackish or salt | Freshwater (can swim in brackish or salt but do not make homes there) | Usually fresh, sometimes brackish or salt | Freshwater | Usually fresh, sometimes brackish or salt |
| **Longevity (years)** | Wild: 3–6 Captivity: 10 | Wild: Usually around 10, possibly up to 20 Captivity: Probably 13–20 | Wild: 3–4 Captivity: 10 | Wild: 8–9 Captivity: Up to 25 | Wild: Up to 10 years Captivity: Unknown |
| **Other distinguishing features** | Large orange teeth and long white whiskers | Slaps tail on water surface | Serpentine tail movement when swimming | Elongated body | Long tubular body with ears that barely project above fur |
| **IUCN Special Conservation Status** | Least Concern | Least Concern | Least Concern | Least Concern | Least Concern |

Sources: Jenkins and Busher, "*Castor canadensis*"; Larivière, "*Mustela vison*"; Larivière and Walton, "*Lontra canadensis*"; Willner et al., "*Ondatra zibethicus*"; Woods et al., "*Myocastor coypus.*"

of brown or gray underfur that is soft and dense with long, coarse guard hairs that are brownish-blond or reddish-brown in color. Males (averaging just under 15 pounds or 6.7 kilograms) are around 5–10 percent larger than females (averaging almost 14 pounds or 6.3 kilograms).[14]

Countless other physical features allow nutria to thrive in their wetland habitat. A few stand out in their importance. One is that nutria are keen smellers. Scientists have found that nutria brains are well developed in the areas that coordinate scent, as prototypical nutria social behaviors like territory marking and family-member recognition rely on excellent scent detection. Oily secretions from scent glands near the mouth and anus are used to comb and waterproof their fur or to groom members of the family group.[15]

Nutria also have pinpoint hearing and tactile abilities. They can detect the movement of predators while in or out of the water, a necessity that comes with having mediocre eyesight. Lack of visual acuity is apparently acceptable in their murky habitat, where excellent vision might not be particularly useful. Instead, nutria have whiskers (also called *vibrissae*) with unusually large numbers of sensory cells at their base that allow them to navigate easily. Another mitigation of poor visual perception is the location of eyes near the top of the head, allowing nutria to accurately spot the movement of predators like alligators (*Alligator mississippiensis*), red foxes (*Vulpes vulpes*), marsh harriers (*Circus aeruginosus*), tawny owls (*Strix aluco*), and water moccasins (*Agkistrodon picivorous*) while they are almost completely submerged underwater. To allow breathing during quick trips to the water's surface or while partially submerged, nostrils are also located on the upper head area.[16]

Nutria have webbed hind feet to help them swim through water more expeditiously than land-dwelling rodents. The first four digits are webbed, and the fifth is free for grooming. To consume underwater plants efficiently, nutria have four large, conspicuous incisors with orange surfaces and 20 teeth overall. Nutria mouths also have an unusual and utilitarian hair-covered palate with a soft cushion that closes the cavity below the teeth to prevent water from running into their mouth while eating.[17]

Finally, nutria can remain submerged underwater for over 10 minutes, using a complex internal system that conserves oxygen by distributing it preferably to the brain. Their red blood cells are unusually large, which probably allows for excellent oxygen retention in the bloodstream. Nutria exhibit a slow heart rate and reduction of blood flow to skin and appendages while diving, and can tolerate physiological consequences of diving like heightened carbon dioxide levels better than most mammals. These features, when combined with a torpedo-like shape, make nutria skilled swimmers even though they lack the salient flat tail exhibited by beavers.[18]

Fur color is another interesting physical feature of nutria. Some fur farmers have observed captive nutria that produce color variants that are different from the typical brownish-blond or reddish-brown guard hairs when selectively bred, owing to a complex series of genetic mutations and inheritance patterns. Variant hues observed in commercially raised nutria include beige, golden, black, and white. Color variation also occurs in the wild among subspecies; *M. c. melanops* of southern Chile has darker fur than *M. c. coypus* from central Chile. This difference probably occurs because darker fur absorbs more heat, a necessary benefit near the tip of South America where temperatures usually do not surpass 50°F (10°C), even during the summer.[19]

## Where Do Nutria Live?

Nutria are native to most areas of South America below 20° latitude (from central Bolivia and southern Brazil to Tierra del Fuego). Four widely recognized subspecies (also called *races* or *genetically distinct populations*) of nutria are found in the southern part of the continent, including *Myocastor coypus bonariensis* (northern Argentina, south-central Bolivia, Paraguay, Uruguay, and extreme southern Brazil), *M. c. coypus* (central Chile), *M. c. santacruzae* (Patagonia), and *M. c. melanops* (restricted to central Chile's Chiloé Island).[20]

Historically, nutria were confined to these areas. But following various introductions to perpetuate the fur industry, escapes and emancipations allowed exotic nutria populations to be established all over the world. Indeed, nutria have been introduced from South America to every continent except Australia and Antarctica; in North America, feral nutria have been reported at one time or another in at least 40 states and three Canadian provinces since their introduction.[21] To the extent of current scientific knowledge, viable and feral populations of nutria are currently found outside of their natural range in at least 16 states, including Louisiana, Mississippi, Alabama, Florida, Georgia, North Carolina, Virginia, Tennessee, Arkansas, Maryland, Delaware, New Jersey, New Mexico, Colorado, Washington, and Oregon. Nutria are also found in the extreme southern regions of Ontario and Nova Scotia in Canada; extreme northeastern Mexico; France, Italy, Germany, the Netherlands, and Belgium; the former Soviet Union, the Middle East, South Korea, southern China, India, and Japan; Kenya and possibly other parts of eastern Africa; and perhaps a few other locations with sporadic sighting histories or records, such as Ireland. In some areas, such as Utah and the midwestern United States, released or escaped nutria did not survive in the wild and failed to establish a substantial breeding population. In other areas, like California and Great Britain, wild nutria were successfully extirpated with control programs.[22]

*M. c. bonariensis* was probably the subspecies most commonly introduced outside of South America, although some findings suggest that nonindigenous North American nutria are of mixed race.[23] Whatever their subspecies, nutria generally multiply quickly when introduced to an area, unless the location has cold weather. Nutria are sensitive to cold temperatures, as their tails contract frostbite and become infected easily, and this reason is the prevailing hypothesis for why nutria have failed to become established in places like the midwestern United States, northern Europe, or areas with high elevations.[24]

## Early Accounts

So how did nutria get their name? Part of the answer depends on whether one refers to the furbearing rodent by the term *nutria* or *coypu*. When Spaniards first came to the Americas, they encountered many unfamiliar flora and fauna, among them the "nutria." They figured the rodent was a type of otter, which was a familiar animal from Spain, so they named it with the Spanish word for otter—*nutria*. In some parts of South America, using the Spanish word *nutria* for the rodent and not otters has stuck. But for the most part, the common name for nutria in Spanish is *coipo* (or, in American Spanish, the word is *coipú*). In this case, Spaniards borrowed and corrupted the word *kóypu*, meaning "water sweeper," from the indigenous Mapudungun people of Patagonia to name the unfamiliar rodent.[25] Convention dictates that the name *nutria* is generally used in North America and Asia, but *coypu* is applied to distinguish the rodent from otters in Spanish-speaking countries. Another unwritten convention is to use *coypu* in reference to the rodent when wild in its natural geographic range, and *nutria* when in the capacity of a furbearer, because nutria were transplanted out of Latin America and into North America. I follow both of these conventions herein.

Perhaps the most famous early account of nutria comes from a book published in 1872 called *Beeton's Brave Tales, Bold Ballads, and Travels and Perils by Land and Sea*. Englishman Samuel Orchart Beeton, who was better known as the first British publisher of *Uncle Tom's Cabin* and a self-help pioneer with a business that produced household-themed magazines, edited the work.[26] Usage of several names for the animal we now call *nutria*, as well as confusion with otters, is apparent from Beeton's prose. His writing also features an excellent physical description of nutria, focusing on the animal in its natural habitat as well as its value to the fur industry:

> The Coypu is a native of Chil[e], whence the skins are largely imported under the name Nutria or American Otter. Both names have the same signification, and both are eminently correct, inasmuch as the Coypu belongs to the rodents and not the weasels. . . . It swims nearly as well as [a beaver], and in like manner guides itself with its tail, which is, however, rounded instead of flat. The chief instruments of propulsion are the hind-feet, which are broadly webbed, and very large. . . . As is the case with most aquatic animals, the Coypu is never completely wet, for although it may have been swimming about for an hour, as soon as it ascends the bank the water rolls off its fur, and leaves the creature quite dry. It is for this purpose that the coat of the Coypu is twofold—first by a thick coat of wool to keep heat in and wet out, and secondly an outer thatching of long hairs over which the water rolls when the animal reaches dry land.[27]

English chemist William Thomas Brande published another well-known early account of "coypú" in his *Dictionary of Science, Literature, and Art* (1852), again beginning with natural history but inevitably turning to the rodent's importance as a furbearer:

> Like the beaver, the coypú is furnished with two kinds of fur; viz. the long ruddy hair, which gives the tone of colour, and the brownish ash-coloured fur at its base, which, like the down of the beaver, is of much importance in hat-making, and the cause of the animal's commercial value.
>
> The habits of the coypú are much like those of most of the other aquatic [r]odent animals. Its principal food, in a state of nature, is vegetable. [The nutria] affects the neighbourhood of water, swims perfectly well, and burrows in the ground. The female brings forth from five to seven at a time; and the young always accompany her. . . .
>
> Nutria fur, largely used in the hat manufacture, has become, within the last fifteen or twenty years, an article of very considerable commercial importance. The imports fluctuate considerably, as many as 600,000 skins having been sometimes imported annually from Buenos A[i]res and Chil[e], but the wars between these states have reduced the exports to about 3,000 skins.[28]

Although the names *nutria* and *coypu* were used widely in formal accounts after the late 1700s, natives of South America employed other monikers for *Myocastor coypus*. Various aboriginal tribes still maintain traditional names for the animal; nutria are called *quiyá* in the language of the Guaraní, an indigenous people from South America's interior, and *caxingui* by the Tupi people of Brazil (where nutria are also known as *ratão-do-banhado*).[29]

Common names for *Myocastor coypus* vary across Europe. In France, the coypu is known as a *ragondin*, while in Italy, the popular name is *nutria* (which defies the aforementioned conventions) but the rodent is also called *castorino* (little beaver) after the Italian common name for its fur.[30] In Netherlands, the nutria (coypu) is known as *beverrat*, which, as the phonetic pronunciation implies, means "beaver rat"; and in German, nutria are known as *wasserrattes*—water rats.[31] Of course, both *beverrat* and *wasserratte* are technically misnomers because nutria are not members of the family Muridae, which includes all of the "true" rats.

During the nineteenth century, European travelers to South America found it entertaining to observe members of native cultures who wore nutria fur and ate roasted nutria meat with a stick.[32] Indeed, aboriginal groups such as the Guaraní and Mapudungun people knew about coypu for hundreds of years. So did members of the ancient Chimú culture that ruled the northern coast of Peru from around 850 to 1470, before Inca conquest of western South America; friezes of repetitious images featuring squirrel-like animals that are probably nutria have been found by anthropologists on the walls of their capital.[33] But despite their iconic status in these ancient cultures, the first formal description of coypu was produced in 1782 by Father Juan Ignacio Molina (Abate Molina), a Chilean Jesuit priest and naturalist. Educated at the Jesuit College in Concepción, Molina was force to leave Chile in 1768 when the Jesuits were expelled. He sought asylum in Bologna, Italy, where he became a professor of natural sciences and wrote his magnum opus, *Saggio sulla Storia Naturale del Chil[e]* (1782), which was the pioneering account of Chile's natural history. The work described many species to science, including coypu, for the first time.[34] Incidentally, it was five years later when Molina proposed a hypothesis for which he is probably more well-known—he suggested that South America was populated from South Asia through the island chains of the Pacific, and that North America was populated from Siberia.[35]

Molina attributed the scientific name *Mus coypus* to nutria, and *Mus* remains the accepted genus for animals commonly referred to as *mice*. But large size relative to other rodents and subtle differences in facial and dental structure eventually proved nutria to be different enough from mice to separate them in the taxonomy. The current genus *Myocastor*, from the Greek *mys* and *kastor* (mouse-beaver), reflected these differences from mice and was assigned by Scottish science writer Robert Kerr in 1792 with his seminal work *The Animal Kingdom*, often regarded as a taxonomic authority on many species.[36] Meanwhile, a French naturalist named Étienne Geoffrey Saint-Hilaire, who campaigned for the now widely discredited "use it or lose it" evolutionary theories of Jean-Baptiste Lamarck and was known for an unorthodox belief that species transmutate in time, experienced similar issues when suggesting a formal name for nutria. His binomial name for nutria, *Myopotamus bonariensis*, never caught on with the scientific community—although *bonariensis* now refers to the northernmost subspecies.[37]

## Life in the Water

One of the most conspicuous characteristics of nutria is that their lives revolve around water. Nutria can live in any area with a high density of vegetation along its banks, including rivers, lakes, sluggish streams, and marshes. Typically known as freshwater dwellers, nutria inhabit both freshwater and brackish (mixed salt and freshwater) habitats in North America. They are found at elevations as high as 3,900 feet (1,190 meters) in the Andes, but for the most part reside in sea-level habitat.[38]

The semiaquatic lifestyle of nutria fascinated nineteenth-century surveyors of Patagonia, who wrote of their interest in how nutria raised young with family groups. For example, naturalist Charles Knight broached the social behavior of nutria with his monograph titled *Sketches in Natural History* (1849). Knight primarily expressed his interest in nutria as wild animals despite the requisite mention of their exploitation as a furbearer:

> This animal [the nutria] is gregarious and aquatic, residing in burrows which it excavates along the banks of rivers; and in these burrows the female produces and rears her young, from three or four to seven in number, to which she manifests great attachment. . . . The coypu remained unknown to the scientific world, while thousands of its skins . . . for more than forty years had been annually imported into Europe for the sake of the fine under-fur, which, like that of the musquash [muskrat] and beaver, is extensively used in the manufacture of hats.[39]

Along the banks of their watery home, nutria use large amounts of local vegetation to construct burrows that are central to their gregarious life.

Life on the water. Nutria sometimes maintain lodges and feeding platforms along the water's edge. Courtesy of Shutterstock, image #294668792.

Burrows may consist of a single, simple tunnel or can extend for 3–20 feet (1–6 meters) with several entrances and an elaborate system of passages or chambers. Complex burrow systems are often expansions of simpler arrangements, and may contain some entrances that are above or below the water.[40]

Burrows are hubs of social activity that are integral to nutria survival. One purpose of burrows is to protect nutria from inclement weather by moderating ambient temperature. Because nutria cannot tolerate freezing temperatures, they increase burrow use during severe winter weather like sleet and wind and spend more time constructing burrows when the temperature dips below 39°F (4°C). Indeed, Argentinean scientists report that burrow temperature may remain between 45–50°F (8–10°C) even if outside temperatures range from 25–75°F (−4–24°C).[41] Burrows also provide protection from predators like alligators, hawks, and snakes. Webbed feet make nutria awkward and vulnerable to predators if they are on land, so burrows provide convenient anti-predator retreats that are in or near the water.

Nutria do not build every burrow they use, and they can sometimes be quite flexible with their sources of cover. Abandoned beaver lodges or muskrat dens are sometimes usurped by nutria, as are farm buildings, straw stacks, and rabbit burrows.[42] These complex burrow systems, lodges, or dens that have been "borrowed" from other animals are sometimes used

Lodged in. A female nutria and her kit near their home, photographed in 1984. Courtesy of Louisiana State University Special Collections: Donald W. Davis Slide Collection, Louisiana Sea Grant Collection Images, Louisiana Digital Libraries.

as full-fledged family quarters. Nutria may birth litters in open nests at the water's edge, but many burrow systems contain a floating platform of veg-etation or a grass-lined den that is used as a nest for kit rearing.[43] Both sexes assist in the construction and maintenance of burrow systems and platforms throughout the entire year, and these areas are usually shared by a male, two to three related females (e.g., sisters, cousins), and their offspring. Females rule the family group for most of the year—they are dominant in social interactions except during breeding, when males dictate social inter-actions more strongly.[44]

Most nutria occupy a small area throughout their lives and travel less than 600 feet (183 meters) daily. The exception is when nutria disperse long distances to find a new home after a disaster like a flood or hurricane, or if introduced into an agricultural setting with few suitable areas to construct burrows. Males that are "resident" to a burrow (or burrow system) defend a territory around it that is about 1,000 feet (305 meters) in length, patrol-ling the area for intruders to defend the nest and their right to breed with nearby females. As young males become sexually mature, the resident male ousts them from the family group via aggressive interactions like chasing or fighting, even if the newly mature males are his sons or close relatives. This means that a young male often travels alone while looking for an area to establish himself as a resident male.[45]

Gregarious living requires nutria to expend energy guarding territory from other nutria. But coloniality also has several advantages. More individuals nearby means more helpers to gather vegetation for building and maintaining the group's burrows. It also ensures that more individuals are available to detect predators and stop family members, especially kits (young nutria that are less than one year old), from leaving a burrow and becoming vulnerable. Males are particularly defensive of burrows after a new litter is produced. Their efforts prevent predation and deter other animals that want to use the living quarters.[46]

Another advantage of colonial living is that group members can huddle together to conserve energy and create warmth during cool mornings. Nutria generally inhabit places where winters are mild, but huddling is nevertheless helpful during less balmy parts of the year, especially because nutria do not hibernate.[47] To make up for time lost during the night that was spent huddling instead of feeding, nutria can become temporarily diurnal. Nutria are usually nocturnal, although during warm months they are sometimes active in the day between periods of sleep to feed, groom, and swim. Generally, nutria are most active at twilight and into the night, but daily activity is highly variable and affected by day length, seasons, and temperature. When the weather is cool, their active period ceases earlier in the night than when the weather is warm; when nutria are active during the day, they enjoy mild warmth (less than 86°F or 30°C) and often sunbathe during the summer. Interestingly, nutria seem to dislike excessive heat. They typically retreat to a burrow or into the water for a swim to cool down if it becomes hot and humid during midday.[48]

These instances offer an unusual opportunity to observe nutria swim in broad daylight. Swimming very low in the water so that they are submerged nearly to the eyes and nostrils, nutria entertain wildlife enthusiasts with their duck-like swimming motion, which involves thrusting their webbed hind feet. Prior to swimming, nutria often remain motionless under sparse vegetation or in their burrows so as to not call attention to a predator. Only their noses and eyes peek above the waterline. Then, when startled (or sometimes for no apparent reason), nutria enter the water with a loud splash and start stroking between vegetation. Because nutria have a sophisticated oxygen-conservation mechanism that allows them to remain submerged underwater for up to 10 minutes, they often swim long distances before resurfacing.[49]

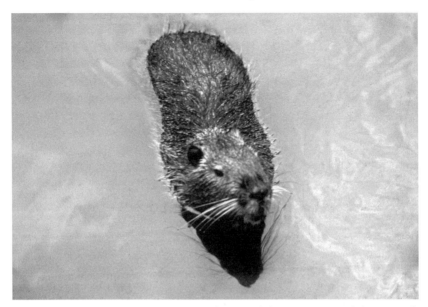

Taking a swim. Nutria are semiaquatic rodents, and their swimming ability is excellent. Courtesy of Louisiana State University Special Collections: Donald W. Davis Slide Collection, Louisiana Sea Grant Collection Images, Louisiana Digital Libraries.

Nutria are almost never seen away from water, and they are usually seen while swimming. When nutria enter and leave the water during swimming trips, they leave a telltale sign of their presence called a *slide*—a foot-wide muddy trail created by the nutria's sliding belly. Nutria also swim with their narrow, pointed tails zigzagging or arching upward in the water behind them. This differs from the round tails of beavers, which are almost never seen while swimming.[50]

Nutria often leave two-inch-long droppings in the water, where up to 86 percent of their feces are produced. Droppings have parallel grooves along their entire length and are usually dark green, brown, or almost black because they contain undigested plant matter. Coprophagy (reingestion of feces to extract additional nutrients) sometimes occurs when nutria return to the nest.[51] Tracks are also diagnostic signs that nutria are nearby. Nutria footprints are often confused with beaver tracks if the beaver's fifth-toe webbing does not print.[52]

Sometimes, nutria can be detected from their unusual vocalizations. Many peaceful nights along Louisiana's shores are interrupted by the shrill cries and piglike grunts of nutria, offering a message not yet known to science and sounding in chorus over the bayou.[53]

## What's for Lunch?

Most of a nutria's waking hours consist of feeding. Nutria are voracious herbivores that feed opportunistically and consume approximately 25 percent of their weight daily. They also excavate soil, finding small food items like bark and roots to hold and manipulate with their forepaws—an otter-like behavior that was probably noticed by the Spanish settlers who gave nutria their misnomer. Most feeding occurs at night, but feeding may occur at any hour when weather is cold and food is not widely available.[54]

Perhaps the most interesting component of nutria feeding behavior is the circular, flattened "feeding platforms" that nutria build in shallow water with coarse, emergent vegetation. Preferring the lower, more succulent portions of plants, nutria often feed by cutting off vegetation near the waterline and carrying it (or swimming with it) to a feeding platform for consumption. Also used as areas for birthing, grooming, or just "hanging out," the 3–6-feet-long (1–2-meters-long) platforms see increased activity during cold weather because they are a warm place for nutria to put their hairless feet. Logs or other floating objects in the water are sometimes used as makeshift platforms as well.[55]

Nutria eat a wide variety of emergent, floating, and submersed plant species throughout their natural and introduced range, and they feed primarily on the water's surface. Seasonal variation in forage quality and availability influences their flexible diet greatly. For example, when food is scarce because green parts of plants are unavailable during winter, nutria also consume roots, rhizomes, tubers, and bark. Nutria will even eat garden crops and lawn grasses that are next to water if it is the only food around.[56]

In Louisiana, more than 60 species of plants are eaten by nutria. A few of nutria's favorite foods include alligator weed (*Alternanthera philoxeriodes*), American cupscale (*Sacciolepis striata*), arrowhead (genus *Sagittaria*, *Nephthytis*, or *Sygonium*), bulrush (family Cyperaceae), cattails (genus *Typha*), common water hyacinth (*Eichhornia crassipes*), cordgrass (genus *Spartina*), pennywort (genus *Hydrocotyle*), rushes (family Juncaceae), reeds, (order Poales), and sawgrass (genus *Cladium*). Alfalfa (*Meidcago sativa*), corn (*Zea mays*), rice (*Oryza sativa*), and sugarcane (genus *Saccharum*) are also fair game for hungry nutria, albeit usually less available.[57]

Nutria have broad tastes for plant life and eat in high volume. These tendencies can rid an ecosystem of its plant life at an alarming rate when nutria are in high densities. The negative effect is exacerbated by nutria wasting around 90 percent of the plant material they forage as they concentrate most of their feeding on stems.[58] Usually this results in vegetation that is reduced to stubbles and piles of clipped vegetation along the water's edge—both are telltale signs that nutria have set up shop. While nutria sometimes eat undesirable aquatic plants and clipping of old vegetation allows for emergence of new flora, the "wasteful" feeding adversely affects wetland ecosystems by denuding otherwise edible vegetation and causing production of less favored foods for nearby wildlife.[59] Thus, nutria have an unseemly reputation and are partly responsible for turnover of rich wetland ecosystems into vast areas of open water.

## Breeding Like Rats

At high densities, nutria reduce the number of emergent plants and can potentially turn wetlands into unproductive waterholes. So how do they reach the high density that brings about this impact? The answer to this question is simple—nutria breed like rats.

Biologists can age nutria based on a combination of molar wear, eruption patterns (when teeth enter the mouth and become visible), body length, body mass, and hind foot length.[60] According to these studies, many wild nutria do not live more than three years, owing to the risk of predation and a sometimes sparse availability of food. But nutria can sometimes live for six years in the wild, and more than 10 years in captive situations such as a fur farm or zoo.[61] During those years, nutria can produce many offspring and may breed during any month of the year. Peak birth periods occur during December–January and June–July in Louisiana, but female nutria are what scientists call *polyestrous*, which means that they are sexually receptive at several different points of the year.[62] Another reason for their high-volume reproduction is that nutria begin reproducing early in their lives. They become sexually mature at around 4–8 months, despite not reaching their full size until about 1.5 years. Nutria fecundity also increases with age. For all of these reasons, female nutria often give birth to 6–10 litters during their lifetime.[63]

Litter size and birth timing are other reasons for the fruitful nature of nutria. After a pregnancy of around 130 days, females usually give birth to 3–6 kits (range: 1–13 kits, mean = 4.73) and are able to breed again only two

Out in the open. Nutria are more susceptible to predators when foraging away from the water or their conspecifics. Courtesy of Louisiana State University Special Collections: Louisiana Department of Wildlife and Fisheries Slide Collection.

Kits and caboodle. A female nutria nurses her kits. Litter size is usually three to six. Courtesy of Edwin Butter/Shutterstock, image #240150688.

days afterward.[64] This means that in a one-year span, females can produce two litters of several kits and already be pregnant with a third. It is therefore no surprise that scientists considering a population of 8,000–11,000 nutria in Great Britain concluded that doubling to 15,000–18,000 could occur within a year if not controlled by hunting or trapping.[65] One nutria population in Maryland exhibited a whopping annual productivity of 8.1 young per female. And in some populations, female nutria have churned out an average of 15 young per year![66]

Nutria replace themselves quickly, but various factors still limit their reproduction. Only about 60 percent of conceived kits are born, and not all birthed kits survive.[67] One reason is the nearly 25 percent of litters that are miscarried or aborted. During inclement weather or lack of food, for instance, embryo reabsorption is not uncommon. This probably prevents female nutria from investing energy into a pregnancy that is unlikely to produce reproductively viable offspring. Scientists have also detected selective abortion of litters that are small and predominantly female, perhaps as a mechanism for controlling the sex or quality of offspring.[68] As for kits that are born, predators like bald eagles (*Haliaeetus leucocephalus*), great horned owls (*Bubo virginianus*), marsh harriers (*Circus aeruginosus*), red foxes (*Vulpes vulpes*), raccoons (*Procyon lotor*), red-shouldered hawks (*Buteo lineatus*), and great blue herons (*Ardea herodias*) are on standby, ready to predate inexperienced young nutria. Adults that stray from the lodge to forage are also susceptible to predation.[69]

Although an omnipresent predation threat exists in their environment, birthed nutria kits are largely independent. Born fully furred and precocial with open eyes, nutria kits are usually swimming and eating plant material within 24 hours of birth. Weighing around half of a pound (225 grams) at birth, kits gain weight quickly during their first months of life and nurse for seven to eight weeks while eating vegetation. With eight teats that are high on her sides, mother nutria can nurse in almost every position, even while lying on her stomach or taking a dip in the water. After another few months, her babies are sexually mature and ready to make more nutria.[70]

## Infections and Parasites

No reliable records on mortality rates from infections and parasites exist for nutria. But in places like Louisiana where nutria population densities are high, these sicknesses can spread quickly and death rates are probably considerable. Across their natural and introduced range, nutria populations often suffer from common parasites like flatworms (phylum Platyhelmin-thes; class Trematoda and Cestoda) and roundworms (phylum Nematoda). Some populations are also susceptible to coccidiosis, equine encephalomy-elitis, leptospirosis, papillomatosis, paratyphoid, rabies, rickettsia, salmonel-losis, sarcosporidiosis, and toxoplasmosis.[71]

Nutria skin is quite susceptible to objects that may penetrate it, so another common issue is chronic irritated skin from the small, one-seeded fruit of the smooth beggartick plant (*Bidens laevis*). Barbed beggartick seeds entangle in nutria fur easily and often start a chain reaction of dermatitis, infection, depression, and loss of appetite when they puncture the skin.[72] Nutria fur or skin can also harbor external parasites such as chewing lice (*Pitrufquenia coypus*), fleas (*Ceratophyllus gallinae*), and ticks (*Dermacentor variabilis, Ixodes arvicolae, I. hexagonus, I. ricinus*, and *I. trianguliceps*).[73] For nutria in captivity that are overcrowded, malnourished, or exposed to poor sanitation, health issues like hepatitis, nephritis, neoplasms, and pneumonia can become recurrent ailments.[74]

Perhaps the most infamous nutria parasite is a roundworm known to the scientific world as *Strongyloides myopotami*, which is usually called "nutria itch" by folks who handle nutria pelts or work and recreate in wetlands. Com-ing from tiny eggs that enter the water via nutria feces, the hatched larvae of *S. myopotami* are often present in nutria-inhabited water. The itch may be responsible for occasional periods of low reproduction and high mortality among nutria, and *S. myopotami* larvae probably cause infections that kill large numbers of susceptible nutria, usually the very old or young members of a population. *S. myopotami* is found in most nutria populations along the Gulf Coast—80–90 percent according to one report—and the effects are pro-found for both humans and nutria. They burrow into nutria skin to grow into adults before restarting their life cycle, but the larvae also cause an aggravat-ing and severe rash for humans who are unfortunate enough to experience skin contact with them because they burrow into a person's skin as if he or she were a nutria. Because humans are not their natural host, *S. myopotami* larvae burrow into human skin until they inevitably die. But the itch, and the agony that goes with it, can go on for weeks.[75] As with all wild animals, people should refrain from physical contact with nutria and take obvious precau-tions if handling them during monitoring or trapping-related activities.

## Trapping and Methods of Study

Kill-trapping is an inexpensive way to obtain an animal for its pelt, but in modern times the practice has become less common and highly controversial. Historically, nutria have been trapped with body-grippers called Conibear traps that are designed to kill animals quickly.[76] The nutria is lured into the trap with bait, or the trap is placed near a burrow entrance. When a trap is triggered, the wires usually close on the nutria's neck, which kills the animal by fracturing the spinal column and closing the trachea and the blood vessels to the brain. Conibear traps that are meant for beaver (#330 or equivalent) or a somewhat smaller animal (#220 or equivalent) are highly effective for trapping nutria, either to acquire the pelt or for control. Wire-cable snares, which are anchored cable or wire nooses that trap nutria about the neck or the body, have also taken many nutria over the years. Leghold traps, which operate in an obvious manner that is consistent with their namesake, can be used to capture nutria but they are less efficient than Conibear traps.[77]

Nowadays, kill-trapping is a less frequent pastime and nutria are usually shot if deemed to be the target of control efforts. Cage-type live traps, where the baited nutria sets off a trigger that closes a door behind it, are the method of choice when nutria are trapped for monitoring or study. Nutria can also be live-trapped on floating rafts or on land, and carrots make the best bait.[78]

Once captured alive, nutria can be anesthetized with ketamine hydrochloride, sodium pentobarbital, or diazepam for data-taking procedures.[79] Nutria are usually marked with tags and punch codes on their ears.[80] Another method for individual identification involves a chemical called Rhodamine B, which can remain in guard hairs for up to 255 days. Some scientists can recognize individual nutria based on their unique whisker pattern.[81]

To monitor nutria activity, scientists may use radio collars, surgically implantable transmitters, or tail-mounted transmitters.[82] Implanted radio transmitters have the advantage of durability because nutria cannot remove them. They also eliminate any possible risk of increased predation, since they are not visible to other animals and do not slow down running nutria.[83] Some scientists attach radio transmitters with beaded collars; others inject "passive negative transponders" (also known as PIT tags) into nutria, which provide information on location while circumventing the aforementioned issues associated with external transmitters.[84] A creative and novel approach to finding nutria comes from wildlife managers in Maryland, where the highly developed scent detection of trained Labrador retrievers is used to sniff out nutria.[85]

Understanding how many nutria are in an area can help wildlife managers determine methods of control or the potential for ecological damage. But it is logistically impossible to count every nutria, and methods of estimating population size typically have either statistical or practical limitations. Thus, nutria populations are often monitored through indirect evidence such as harvest records, and conclusions from these data are generally drawn for a specific area instead of an entire region.

Implicit in the conclusions from harvest records is an unlikely-to-apply assumption that the efforts of the public to trap nutria was equal in all years. Various methods of additional survey are therefore useful in providing an estimate of nutria population changes in a fixed area. For example, biologists sometimes use night counts (when nutria are usually more visible than during the day), active trail counts, and scat counts.[86] While the exact correlation between these surveys and the actual number of nutria remains unknown, these methods produce a general trend of whether nutria populations are increasing or decreasing.

Mark-recapture studies can be used to obtain population density estimates or generate data on population trends. But this method is labor intensive, time consuming, and not entirely foolproof. Calculations for mark-recapture studies assume that a population is "closed," with fixed boundaries and no individuals immigrating or emigrating. Mark-recapture methods also assume that all nutria have an equal chance of being caught, which is very unlikely. Biologists attempt to limit these issues by trapping and retrapping within a short amount of time, usually 8–12 days. Usually data from a combination of these methods are integrated to form a comprehensive picture of nutria population trends so that the limitations of any one calculation are less problematic.[87]

Microsatellite DNA markers—data from repetitive sequences of DNA that have a high mutation and diversity rate in a population—provide a sophisticated (albeit expensive) way to identify and monitor nutria. Because every nutria has different forms of genes in various combinations, using DNA can reveal detailed information about nutria population dynamics and breeding structure.[88]

### Biology and Wildlife Management

Because they are regarded as pests, nutria have been killed or sterilized using every method imaginable over the past few decades. In 2002, Louisiana initiated efforts to combat wetland loss from nutria overpopulation that included a four- to five-dollar-per-tail bounty program.[89] Other states suffering from nutria-induced wetland damage also implemented control protocols around this time. For example, Maryland's Chesapeake Bay Eradication Project (MDCBEP) combined trapping with sterilization and complex detection methods. Oregon relaxed the laws on trapping and shooting nutria outside of city limits and added provisions that disallowed relocation of trapped nutria. Internationally, countries like New Zealand classified nutria as a "prohibited new organism," making their importation illegal.[90]

Although wetland loss in Louisiana has been reduced by 90 percent since initiation of the bounty program, nutria are still chomping away at aquatic flora and necessitating continued efforts to address their presence in the Gulf Coast region. Three questions follow from these developments. First, how do nutria damage wetlands? Second, why is nutria damage to wetlands so unacceptable? And finally, how did nutria evolve from an introduced species relegated to captive fur farms into one of the most notorious invasive species of our time?

The nutria invasion started with a combination of intentional introductions by fur farmers and government officials to provide trappers with more fur resources or to control problem plants. Some accidental introductions occurred after nutria escaped from enclosures. The fecund rodents multiplied so quickly that by the 1950s, 20 million nutria were munching on Louisiana swamps and marshes.[91] Damage to sugarcane and rice fields increased and the state legislature responded by offering a 25-cent bounty for every nutria killed in 16 parishes (a designation equivalent to what most states call "counties").[92] This brought relief from nutria for a decade or two, but attitudes about wearing fur changed and demand for nutria fur decreased. With few natural predators other than American alligators (*Alligator mississipiensis*), which currently warrant no official conservation concern but were at one time considered endangered, the vegetation-clearing nutria became an increasingly salient nuisance.[93] By the early 1980s, complaints shifted from agriculturalists worried about their sugar and rice fields to land managers and concerned citizens worried about wetland loss.[94]

Concern stemmed from the environmental and economic importance of unique wetland ecosystems. A *wetland* is land area that is saturated with water and has distinct ecological characteristics. In Louisiana and the Gulf Coast region, the Franco-English term *bayou* (from the Choctaw *bayuk*) is often used to describe a body of water in a flat, low-lying area, such as a slow-moving river, marshy lake, or wetland.[95] Wetlands are distinguished from other bodies of land or water by their characteristic vegetation, which results from unique soil conditions known to scientists as *hydric*. This means that soil is saturated with water long enough to develop an anaerobic condition—that is, an oxygenless status. Most soils are aerobic (with oxygen), and this allows plant roots to respire. Thus, carbon is released from aerobic soil in the form of carbon dioxide as part of a cycle that keeps energy flowing through the ecosystem. Specifically, carbon dioxide is taken from the atmosphere and converted into organic compounds like sugar during photosynthesis in plants; these compounds are either stored in the plant and eaten by animals, used by plants as structural components, or respired by plants to release more energy and carbon dioxide.[96]

But in an anaerobic condition where oxygen is not widely available, intense competition occurs between plants for the remaining oxygen, and unique biochemical reactions or plant adaptations result. Most notably, plants found in anaerobic, hydric soils like cattails, sedges, and water lilies have structures called aerenchyma, which are internal spaces in stems that allow for transportation of atmospheric oxygen into roots.[97] In fact, four major types of wetlands are delineated mostly by a particular type of vegetation: swamps, which are forested or have shrubs; bogs, which are covered by mosses or dead plant material; marshes, which feature herbaceous instead of woody species; and fens, which have high dissolved mineral levels but few other plant nutrients. No such categories exist for the water covering the ground, which can be freshwater, saltwater, or a combination of the two called brackish.[98]

Wetlands cover only 5 percent of the land surface in the contiguous 48 states, but they support about a third of the plant species.[99] This means that wetlands are key reservoirs of biodiversity because they provide resources, shelter, and nesting sites for wildlife. Species that depend on wetland habitats for their survival include over 50,000 species of insects and invertebrates that eat the algae common in wetlands; fish and amphibians that need the protection of natural hollows and the constant presence of water provided by wetlands for nursery and hatchery grounds; migratory birds that eat native insects on their way to and from breeding locations; and rare, unusual, or endemic plants like baldcypress (*Taxodium distichum*), tupelo gum (*Nyssa aquatica*), dwarf palmetto (*Sabal minor*), wax myrtle (genus *Myrica*), pitcher plants (family Nepenthaceae or Sarraceniaceae), eelgrass (genus *Vallisneria*), and ditch grass (genus *Ruppia*).[100]

The environmental and aesthetic importance of wetlands speaks to their economic worth. Three-quarters of the commercial fish (and shellfish) stocks in the United States depend solely on estuaries (coastal wetlands with brackish water that are fed by a river and connected to the open sea) to survive. Commercial fishing of shrimp, oysters, and blue crabs that rely on the coastal wetlands as a nursery for their young accounts for several million dollars of Louisiana's economy. Louisiana also sells a few hundred thousand hunting licenses and almost a million fishing licenses annually to folks who depend on the wetlands as a habitat for their game.[101]

Wetland function extends to other societally important needs such as flood control, groundwater replenishment, oil and gas production, shoreline stabilization, storm protection, and water purification. For example, porous sediments in wetlands allow water to filter down through soil and overlying rock into aquifers that supply us with drinking water. Wetlands can serve as "recharge" areas when surrounding water levels are too low and "discharge" zones when levels are too high. During storms, wetlands are a buffer between sea and land that reduces the speed and height of waves or floodwaters. Wetlands also have plants that can absorb and filter heavy metals from drinking water, and they feature organisms and biofilters that can remove toxic substances such as pesticides and industrial discharges. Ecotourism is another economic benefit, as visitors to wetlands enjoy fishing, boating, swimming, camping, hiking, wildlife watching, and the "swamp tours" run by local outfits. And commonly used materials that can be extracted from wetlands include salt, fuelwood, fibers, and dyes.[102]

Most American wetlands are found in the Gulf Coast region, and 40–45 percent of wetlands in the 48 contiguous states are found in southern Louisiana.[103] As the drainage gateway to the Gulf of Mexico, the Lower Mississippi Regional Watershed (LMRW) drains more than 24 million acres (97,000 km²) in seven states along the Mississippi River.[104] Thus, the sustainability and ecological balance of Gulf Coast wetlands is a national issue. Unfortunately, wetlands all over the country are disappearing quickly, and reports indicate that Louisiana loses an area of wetlands about the size of a football field every hour.[105] Current threats to wetlands include development and urbanization, disturbance from geological exploration, coastal erosion and rising sea levels from hurricanes, large-scale logging, and coast subsidence from exploitation of oil and gas resources. After their introduction to the region, nutria became another major contributor to wetland damage.[106]

Because wetlands are ecologically important, efforts to save them are numerous and varied. These efforts range from restoration of native grasses, to limiting development, to cleaning up damage from oil spills, to nutria trapping or bounty programs. Central to the rationale for nutria-related initiatives is the labeling of nutria as an *invasive species* and understanding what this term means. Definitions vary, but most scientists consider an invasive species to be one whose introduction does or is likely to cause harm to the environment, economy, or human health.[107] Several terms like *alien*, *nonindigenous*, *nonnative*, and *exotic* are used more or less interchangeably for species that exist outside of their natural range; *introduced* implies that individuals were moved from one location to another via human activity.[108]

Not all introduced species are invasive. In fact, most are not. For an introduced species to become invasive, it must be transported outside of its natural range beyond geographic barriers like oceans, mountains, and intolerable climate zones. Then it must establish a self-sustaining population. Because introduced populations typically involve only a small number of species members in a new area, they can be highly susceptible to extinction. In the event that a breeding population becomes established, it usually remains highly localized unless members of the population can surmount additional geographic barriers and expand their range. Even then, the introduced species must have the potential to inflict environmental or economic harm, and this tendency does not necessarily manifest because of range expansion.[109]

Nutria qualify as an invasive species because they have jumped through all of these environmental hurdles. As with many invasive species, nutria initially seemed like a benign newcomer. Some were even released purposely to control noxious weeds, and it was a few decades before nutria emerged as major pests that were transforming wetlands into open water.[110] Understanding the fundamental questions of how and why nutria caused so much damage to wetlands is easy after a thorough review of their biology—nutria have a natural inclination to warm, wet ecosystems and multiply quickly. These qualities made vegetation loss imminent after nutria populations were left unharvested. And it is also easy to see why damage to environmentally and economically crucial ecosystems like wetlands cannot be tolerated.

So how did nutria transform from a relatively innocuous introduced species into one of the most notorious invasive species of our time? And why did America's "Third Coast" become ground zero for the invasion? We can attribute some of these developments to the prolific reproduction and voracious appetite of nutria. But to fully answer these questions, we must examine the original reasons that nutria reached the Gulf Coast and other parts of North America, and how an international fur industry built by European settlers preceded nutria introductions. And so we turn to the history of a fur industry that is tightly associated with the Mississippi River and once depended on another large, furbearing mammal with a propensity to eat plant matter—the beaver.

# Chapter 2

# Rat Race

The history of North American expansion might almost be written in terms of the fur trade. Europeans were early attracted to the North American coast by the hope of reaping profits from this trade, and after the beginning of settlement revenue from it was the principal means of sustenance to the early English, French, and Dutch colonies. As European civilization advanced across the continent, everywhere it was preceded by the fur trader, who by the very law of his being operated upon the frontier. Many a nameless trader, intent only upon his trade and caring nothing for the name of discoverer, has been the first white man to set foot upon lands credit for the discovery of which has gone to others. . . . Before him was the wilderness; behind him, over paths he himself had made, poured in an ever advancing tide of settlement.

—**Arthur H. Buffinton**, in a paper presented to the Colonial Society of Massachusetts by Samuel Eliot Morison in January 1916

*In the summer of 1611, a small boat drifted aimlessly between the ice of a large bay north of present-day Canada, occupied by a captain, his son, and seven others with few supplies. Their fate remains unknown and the men were never seen again, probably succumbing to exposure as they meandered through the frozen waters.*

*The marooned captain's name was Henry Hudson, and his mutineers were fresh off a voyage around the west coast of Greenland on a ship aptly named* the Discovery, *which did not go as planned. The casus belli came when the* Discovery *became trapped in ice that winter. As tensions between the sailors grew, they survived onshore with minimal supplies at the southern tip of the bay, waiting for the ice to clear. When it did, Hudson sought to continue west and explore the area more. But the homesick seamen would have none of it, sending him to a certain death in the bay that is now, thanks to British explorers and colonists, his namesake—Hudson Bay.*

*Hudson died that summer, but the effects of his exploration lived on. Through his mission to find a Northwest Passage to Asia via the Arctic Circle, Hudson accidentally found the wide river that travels upstate from modern New York City and thereby laid the foundation for Dutch colonization of the region. It was during exploration of the Hudson River that he traded with Native Americans and obtained furs, and his voyage was subsequently used to launch Dutch claims to the area, particularly the fur trade that emanated from a trading post established at Albany in 1614. The fateful voyage on the* Discovery, *sailed under the British Union Jack, was perhaps Hudson's most influential. Hudson Bay provided access to parts of Canada and the Arctic that were once thought to be landlocked, and resulting British interests in the area's fur trade led to more than two centuries of lucrative business by the Hudson's Bay Company, which at one point became the largest landowner in the world.*

*The resultant network of trading posts formed the basis for western settlement and official European authority in North America, with competition between the Hudson's Bay Company and rival companies that challenged British business interests, like the American Fur Company, coming to influence the present international boundaries of the continent. From Canada, to the American colonies, to the Louisiana Territory, whoever dictated the fur trade was rewarded with empire and fortune, and trappers trailblazed everywhere in search of furbearing animals, ready to make a quick wage along the trade routes.*[1]

✦ ✦ ✦

On a humid autumn evening in 2015, I nonchalantly flipped on the television in my New Orleans hotel room and began preparing for the next day's research in the library. A most appropriate program emanated from the screen. In the infamous episode number 142 of *Seinfeld*, the title character's ex-girlfriend Elaine acts as Peterman Catalog president and charges expensive purchases to her business account, including a new $8,000 Russian sable hat for her neurotic acquaintance, George Costanza. He purposely "loses" the hat in a women's apartment as an ill-advised ploy to return and ask her for a second date, and when Elaine tries to retrieve the hat after she is unable to justify it as business expense, the woman insists it is not in her possession. Trying to save her job, Elaine buys a replacement from a street dealer in Battery Park—"the difference is negligible," says Jerry Sein-feld—but the Peterman accountant shrewdly detects the substitute "sable" hat, which is actually made of nutria fur. "That's a rat hat?" she says, coyly and with characteristic Seinfeldian cadence. "And a poorly made one," says the accountant, "even by rat hat standards."[2]

In their humor, the writers of *Seinfeld* inadvertently highlighted the American prevalence of fur hats as a status symbol and the use of nutria fur to satisfy niche or "replacement" markets. The "rat hat" line made nutria recognizable to many folks across America, but it quickly became clear dur-ing my research that along the Gulf Coast, nutria were already well-known. Indeed, their introduction to the region as a source of fur was legendary and embedded into local folklore.

Louisiana was once home to an omnipresent fur industry that pro-vided families with their livelihoods and generated millions of dollars for the state's economy. Nowadays, fur trappers are much less common, with financial return on furs decreasing in the wake of deteriorating marsh hab-itat and anti-fur activism. The voices of their descendants can be heard in roadside seafood houses, the streets of New Iberia, or in houseboats along the Atchafalaya River. *Trapping was a way of life down here,* they remem-ber.[3] *Can't trap for a living now. . . . Twenty years ago there was money in it, but now you're lucky to break even. My sons will probably be the last genera-tion of trappers.*

*Things just aren't what they used to be.*

While writing this book, I spoke with people who had diverse points of view regarding the international fur industry. I would like to say that I was open-minded toward folks who either sponsored the current sale of animal fur or found nothing wrong with the past intensity of hunting furbearers, but such a statement would be self-serving. To be honest, my background as a naturalist often made me uncomfortable with these philosophies. It was apparent that pro-fur individuals developed their ideology under different circumstances than mine, and the idea of killing animals for any reason, especially for their fur, remains distasteful to me.

But over the course of my research, I began to realize that the discus-sion about how we arrived at our present-day issue with nutria, and how to handle the problem, was not black and white. For example, folks may hunt nutria for fur, for meat, or with the goal of extermination. Because nutria are contributing to coastline destruction, their argument was that killing nutria helped the environment and provided supplementary income for commercial fishing crews whose shrimping businesses were hit hard by the area's environmental issues and a poor market. It was a position that came from a different perspective than mine, but one that I understood objectively—even though I could not stomach trapping or shooting nutria, or any animal, myself.

In another even more intriguing category were individuals who shoot or trap nutria or other furbearers not only for practical or economic reasons, but also because it preserves what they consider to be cultural tradition. These folks have no disdain for the fur trade of years past that nearly drove the beaver to extinction; decimated bobcats, otters, and muskrats; and eventually led to the introduction of nutria to the Gulf Coast region. They displayed their furs and traps proudly, speaking of a close-knit community in the Barataria region. Their folklore included the best way to skillfully skin "swamp rats," the most efficient methods for laying traps in water, father-son good times in the field, and shop talk of traps, guns, or knowledge passed from generation to generation. To these folks, participating in the modern fur trade was a chance to return to appreciate nature and support a traditional way of life. And if the trapped animals were nutria, then so much the better—as far as these folks were concerned, they were coincidentally helping to save the Gulf Coast's wetlands.

It would be easy to blame the fur trappers of yesteryear for their role in destroying our nation's wildlife or obliterating Native American traditions, and convenient to label today's trappers as rednecks for continuing this controversial practice. But economic circumstances were different in the 1700s and 1800s, and as hard as these attitudes are for me to reconcile, we must all wonder whether we would have conducted ourselves the same way if we had lived during those times. On the other hand, I also wonder why we did not stop after the beaver and some other furbearers were annihilated, but some might say that this no longer matters. Perhaps nature writer Barry Lopez said it best in his epic book *Of Wolves and Men*: "You want to say there never should have been a killing, but you don't know what to put in its place."[4]

My research questions involved the historical and environmental context of nutria introduction to the Gulf Coast region. Why was the annihilation of furbearing mammals on North America during the 1700s and 1800s so systematic that it created desperation for another source of fur? Why did fur farmers then hedge their bets with the introduction of a South American rodent whose biology was relatively unknown? How did fur trappers arrive in Louisiana, and why is the tradition of trapping and hunting there so strong that it once demanded the introduction of nutria? And how did circumstances within the international fur industry change such that they encouraged entrepreneurs to start nutria farms? The answers lie in the events of world history, and the tales of the British, Frenchmen, Spaniards,

Creoles, Cajuns, cowboys, and Indians who made the United States the country it is today.[5]

## Mad Hatters

"Take some more tea," the March Hare said to Alice, very earnestly.

"I've had nothing yet," Alice replied in an offended tone, "so I can't take more."

"You mean you can't take less," said the Hatter: "it's very easy to take more than nothing."

—Lewis Carroll, *Alice's Adventures in Wonderland*

Prior to the arrival of nutria in Louisiana, and before European settlers came to the "New World" and recklessly trapped furbearers until hardly any were left, Native American tribes practiced trapping to obtain food and warm clothing. Depending on the culture and location of the tribe, the choice of game could include land-dwelling furbearers such as mink, rabbits, grouse, bobcats, otters, deer, caribou, and buffalo, or sea mammals such as whales, porpoises, walruses, and seals. Methods of harvest also varied according to the culture, as skilled hunting with bows and arrows or lances was often practiced. Trapping developed as a technological advance over hunting, and North American Indians began using snares, nets, and deadfalls to catch and kill furbearing animals.[6]

Regardless of whether the animal was caught via hunting or trapping, its pelt was used to make clothing, usually after the meat was eaten. For instance, Algonquian-speaking peoples of subarctic North America hunted moose, caribou, bears, and lynx, which were used to make moccasins, shirts, robes, and dresses. Tribes residing in the present-day Southeast United States made a variety of cloaks and shawls for ceremonies or to keep warm during the winter. Bobcats were trapped by Navajo Indians, with the skins used for caps, mittens, or arrow cases and the claws used to decorate clothing; Indians of the Columbia River also used bobcat pelts for making robes.[7]

Living in a region that was mostly devoid of vegetation and where protection from cold was a necessity, aboriginal groups from artic latitudes like the Innu, Inuit, and Yuit were particularly dependent on wild animals for food and clothing.[8] According to a Montagnais (subdivision of Innu) Indian in 1634:

The beaver makes everything perfectly well, it makes us kettles, swords, knives, bread; in short, it makes everything . . . without the trouble of cultivating the ground. The English have no sense; they give us twenty knives . . . for one beaver skin.[9]

Such practices were not limited to cold-weather areas. Even tribes dwelling in balmy weather, such as the Tohono O'odham in modern-day Arizona and the Muscogee in what is now Alabama and Georgia, hunted or fished to acquire skins for clothing and to supplement their otherwise starchy diet with protein.[10]

Ethnologists report consistently that the relationship between native cultures and their kill was spiritual in addition to being practical. American Indians believe that wildlife and natural phenomena are part of a community to which they belong, and various cultures honored the necessity of a taken animal to their survival with rituals or ceremonies. The most famous example is Plains Indians like the Sioux, Cheyenne, Pawnee, and Omaha, who conducted a traditional festival that coincided with the return of the buffalo herds. The gathering featured a "buffalo dance" including men wearing buffalo skins and a great feast. They believed that the dance honored interrelationships of the tribes with the buffalo, celebrated their oneness with buffalo and the whole of nature, and ensured the herd's return each year.[11] Another example comes from some other tribes living in forested areas, who conducted ceremonies to honor their reliance on the bear. The events included rituals such as addressing the bear with honorific titles, or making a conciliatory statement to the bear before or after killing it. By apologizing for the kill's necessity and treating the bear with respect, the hunter was believed to appease the bear's spiritual "keeper" and ensure a steady supply of bears to hunt or trap.[12]

Caring little about traditional attitudes, Europeans saw wildlife as a resource to be exploited and nothing more, with colonial trappers viewing the "New World" as a nearly limitless supply of resources. Given their traditional conflict with this philosophy, it is ironic that Native Americans played such a crucial role in the logistics of the postsettlement fur trade. At first, Native Americans sought more pelts simply to trade for European goods. But as competition between European companies increased, so did the need for Native Americans to guide traders in a never-ending quest for more pelts, and to trap, prepare, and transport those pelts to Europeans for trade. This resulted in a higher price paid to the tribes for harvesting fur, meaning more incentive to increase harvests and an inevitable breakdown of the traditional human-animal relationship. Indian traders responded with increased harvests as prices rose, and anyone who conserved furbearers as a natural resource lost out economically. Thus, the concern of sustainability among many Aboriginals with regard to fur resources waned, and settlers were happy to reap the benefits of this compromise in philosophy.[13]

By the early seventeenth century, the fur trade was quickly becoming a mechanism for European advancement into North America. Initially, European explorers thought little of beaver skins exchanged by the natives they encountered, as they were focused on finding gold and spices in China. But when North Atlantic cod fishermen and sealers in Newfoundland developed a casual but profitable side trade of furs, the word spread quickly up and down the Saint Lawrence seaway. French settlers quickly established a comprehensive network with trade stations near the river.

Indeed, French colonization of North America began in present-day Maine and Nova Scotia during 1604 and was started largely to trade furs with natives. After his exploration led to these settlements, Samuel de Champlain became "Father of New France" in 1608 by founding the territory's eventual capital as a trading post at what would become the city of Quebec. Continuing his exploration westward to Lake Huron, Champlain established a route to the continent's interior and the fur trade expanded quickly. Algonquin and Huron Indians became allies of the French and were soon trapping furbearers voraciously for trade; some French colonists, referred to as *coureurs de bois*, trapped and hunted independently.

All furs were sought for coats, gloves, and other clothing or decoration, but demand for beavers was particularly high. The reason was that beaver hats became fashionable across most of Europe in the mid-1500s. Beaver fur was soft, resilient, and so easily pliable that it could produce many different hat shapes, the most famous being the familiar top hat.[14]

Hats of certain shapes or colors became reflections of socioeconomic identity in European social circles, sometimes indicating the profession, wealth, political party, religious affiliation, or social rank of the wearer. One example among many comes from ecclesiastical heraldry, where a red and wide-brimmed hat designated the wearer as a cardinal, and interactions with the individual therefore required a specific social protocol. The expense of beaver hats made them a particularly strong visual statement about wealth and class, and they became so popular and desirable that by the seventeenth century, the European beaver (also known as the Eurasian beaver, *Castor fiber*) was rather scarce.[15] With the decline of *C. fiber*, settlers came to view the American beaver as necessary to fill an economic void in "Mother Europe." Thus, beavers went from being a source of food and clothing for Native Americans to something that could be exchanged for European goods.[16]

The French and Indian fur trade flourished, but the wealth of beaver meant that other western European empires were also exchanging pelts.[17] In a small Dutch ship called the *Half Moon*, Henry Hudson of the Dutch East Indian Company, a chartered company established in 1602 with a 21-year deal to trade exclusively in Asia, traveled 150 miles up the "Manna-hata" River, which later bore his name. Hudson traveled about as far north as modern-day Albany, then called Fort Orange.[18] By 1609, the Dutch mapped the area and started a furtrading post on Manhattan Island, at the southern mouth of the river. They began trade with the Iroquois, shipped hundreds of beaver and otter skins back to Europe, and were a permanent presence in "New Netherland" by 1624. According to a member of Hudson's Crew:

> The people of the country [likely Mohawk Indians, a subsection of the Iroquois Confederacy] came flocking aboard and gave us grapes and Pompions [*sic*; pumpkins] which we bought for trifles. And many brought us beaver skins, and otter skins, which we bought for beads, knives, and hatchets.[19]

About two years later, Dutch colonists acquired Manhattan from the native Indians, and the city of "New Amsterdam" was formally incorporated in 1653.[20] But Dutch rule of the area was short-lived. The English conquered New Netherland in 1664 and renamed the colony "New York" for the English Duke of York and Albany, who eventually became King James II. It was briefly retaken and renamed "New Orange" by the Dutch a decade later, but the Treaty of Westminster permanently ceded New York to the English in 1674.[21]

Meanwhile, the Iroquois continued bringing furs to Albany, which became a center for English trade along the Hudson River. It was a major development for England, which had been slower to start a fur trade than the Dutch, with Jamestown, founded in 1607, basing its economy on tobacco. A fur-trading post established in Maine by the Plymouth Company in 1607–8 lasted only a year, but the Pilgrims, who founded Plymouth in 1620, were financed by money-conscious investors in foreign ventures who expected the colonists to build fur, fish, and timber operations. It was not long before the *Mayflower* colonists were successful in that endeavor, especially with regard to furs. As historian and writer James Truslow Adams noted in his Pulitzer Prize–winning book *The Founding of New England*:

> There had been a moderate degree of comfort, as well as abundance of food, at Plymouth; so that in September, under the guidance of Squanto [a Patuxet Indian who assisted the Pilgrims after their first winter], the pilgrims

undertook their first trading voyage, sailing to Massachusetts Bay. Their plan had been to explore the country and to make peace with the Indians of that district, as well as to 'procure their truck.' Although gone only four days, the little part of thirteen . . . were eminently successful in all three objects, making the first beginning, on any large scale, of that trade which was to prove their financial salvation. In fact the Bible and the beaver were the two mainstays of the young colony. The former saved its morale, and the latter paid its bills, and the rodent's share was a large one. The original foundations of New York, New England, and Canada all rest with the Indian [fur] trade, in which the item of beaver skins [sic] was by far the most important and lucrative.[22]

Historians estimate that colonists took 12 million pelts of various mammal species during the seventeenth century.[23] But the English would have to outcompete the French if they expected to dominate the beaver pelt trade. Doing so started further north in Hudson's Bay, and further west, where the mighty Saint Lawrence River ran into the Great Lakes.[24]

## Rule Britannia

*When Britain first, at Heaven's command*
*Arose from out the azure main;*
*This was the charter of the land,*
*And guardian angels sang this strain:*

*"Rule, Britannia! rule the waves:*
*"Britons never will be slaves."*[25]

For most of the sixteenth and seventeenth centuries, France commanded the beaver trade with England lagging behind. *Coureurs de bois* and French pioneers like Champlain trapped aggressively throughout Nova Scotia, the Gaspe Peninsula, and other parts of modern-day Quebec. French missionary activity and trade with aboriginals flourished after 1625, with over 10,000 settlers ensconced in Canada by 1690.[26] Using licenses to lease the use of their posts, the French incentivized extension of fur trading and moved into the Great Lakes region and the upper Missouri River. Many pelts were trapped by aboriginal tribes, but as French settled North America's well-stocked interior, they were able to slowly cut out the tribal middlemen.

Southern expansion of the French fur trade occurred when the upper Mississippi was claimed for France by Sieur Duluth in 1679. Michilimackinac,

Vast claim. René-Robert Cavelier, sieur de La Salle (date unknown), French explorer in Louisiana in the seventeenth century. Born in Rouen on November 22, 1648, he explored the Mississippi River and claimed the entire lands that drained into the river for France. Courtesy of State Library of Louisiana.

now referred to as Mackinac Island, quickly became a thriving fur trade center, and Detroit was founded as a trading post in 1701 to cut off the Iroquois and English colonists from the western Great Lakes. Military officer Sieur de La Vérendrye and his sons continued the growth by establishing trading posts at Lake of the Woods and Lake Winnipeg, and other Frenchmen established posts at Lac des Prairies and Lake Nipigon. The result was direct competition near English ports, with aboriginal tribes having multiple choices for selling their goods.[27]

Meanwhile, Jesuit missionary Father Jacques Marquette, who founded Michigan's first European settlement of Sault Ste. Marie, and Louis Joliet, a fur trader, had reached the northern Mississippi River in 1673. Explorer René-Robert Cavelier, sieur de La Salle, followed them a few years later, becoming the first European to travel from the Great Lakes to the Gulf of Mexico. In 1682, La Salle claimed all of the land that drained into the Mississippi River for France, naming the claimed territory *Louisiana* after King

Founder. Pierre Le Moyne d'Iberville, generally regarded as the founder of Louisiana, in the late 1690s. Courtesy of State Library of Louisiana.

Louis XIV. La Salle planned to start a colony and monopolize the region's fur trade for his homeland.[28] The area La Salle claimed turned out to be vast, so his idea was to build a great empire by linking the Saint Lawrence and Mississippi basins. Doing so would trap the English on the Atlantic coast. But because Spain also laid claim to the Gulf Coast, albeit without occupation, the French minister for naval affairs and colonies, Louis Phélypeaux, comte de Pontchartrain, gave Pierre Le Moyne d'Iberville, generally credited with founding Louisiana, the task of locating the mouth of the Mississippi River.

Despite the myriad species of wildlife in the Louisiana Territory, using domestic stock for trade was not a serious consideration. D'Iberville looked instead to redirect beaver and bison pelts from the continent's interior to the lower Mississippi delta. He thought that native traders in the upper valley could be convinced to eschew dangerous, overland travel to meet English traders by offering comparably priced French goods at convenient stations upriver. Once purchased, these furs would be taken to the delta on *bateaux plats*—flatboats—and transferred to oceangoing vessels for shipment to France.[29]

This seemed like an appropriate game plan until residents of New France complained to Paris about Quebecois merchants being "financially ruined" by the upstart sister colony in Louisiana. After being reassigned away from Louisiana, d'Iberville died of yellow fever in Havana, unable to realize his vision for the Louisiana Territory. Instead, d'Iberville's brother, Jean-Baptiste Le Moyne de Bienville, established "Nouvelle Orleans" near the mouth of the Mississippi River in 1718, and the settlement's location achieved its intended effect. The settlement cut off the British from the area and provided the French with a warm-water port that increased profitability by enabling transatlantic fur shipments. With New Orleans, the French plans for a conglomerate Atlantic Coast–Great Lakes–Mississippi River trade route seemed to be realized.

But the French would soon face unrelenting competition from their British archrivals. Two earlier French explorers of the West, Pierre Radisson and Médard Chouart, sieur des Groseilliers, objected to sharing their fur trade profits with Canada's governor and defected to enter English service. In 1670, the pair helped a group of merchant adventurers start the Hudson's Bay Company (HBC), and King Charles II of England granted the HBC exclusive trading rights to the entire Hudson Bay watershed, also called "Rupert's Land," a 1.5 million square mile (3.9 million square kilometer) territory named after Prince Rupert, the company's first governor.[30]

The company's traders and trappers forged relationships with the aboriginal Cree and Assiniboine, creating a vast network of trading posts to deal in furs acquired mostly by First Nations trappers.[31] HBC's furs were among the finest in the world, and the company became the most successful fur-trading entity in history, even functioning as the de facto government in parts of presettlement North America. At one point, the HBC was the largest landowner in the world, with its area covering 15 percent of North America.[32] Thus, the British monopoly thrived, in spite of constant conflict with the nearby French. Skirmishes like King William's War (1689–97), Queen Anne's War (1701–13), and King George's War (1744–48) arose partly because control over Rupert's Land and access to Native American traders was essential for westward expansion of the fur trade.[33] Finally, after border conflicts between the British and French in the Ohio Valley, the perpetually-in-conflict nations declared war in 1756. And so began the French and Indian War, the North American phase of the worldwide Seven Years' War.[34]

The war proved to be a tough road for the French. At the start of the conflict, French North American colonies had a population of roughly 60,000 settlers, compared to two million in the British colonies. Thus, the outnumbered French depended largely on allied Indian forces.[35] In the end, this makeshift coalition was no match for the British. In 1758, the British seized territory east of Fort Duquesne, now Pittsburgh; then, in 1759–60, the British advanced to the Saint Lawrence River and took Quebec and Montreal, making the all-important Great Lakes region accessible to London. In a nutshell, the British won, the French lost, and the ramifications of this outcome continue to define North America.

Under the Treaty of Paris in 1763, France was essentially eliminated as a North American power, as all French possessions in Canada and eastern Louisiana were ceded to Britain. New Orleans and western Louisiana were ceded to Spain, a French ally, in compensation for Spain's loss of Florida to Britain. As insult met injury, the military defeat and financial burden incurred from the wars in North America contributed to civil discord and helped spur the French Revolution in 1789.[36]

For Britain, the Seven Years' War nearly doubled the country's national debt, leading the Crown to impose new taxes on its colonies. Founding fathers of the United States like George Washington, Patrick Henry, and John Hancock were less than happy to comply, and umbrage to "taxation without representation" eventually led to the American Revolutionary War, complete with the symmetry of a French return to North America in 1778 as part of a Franco-American alliance against Great Britain.

As for the HBC, its vast network of trading posts remained the de facto authority in areas of North America, but their empire did not remain unchallenged. During the late 1700s and early 1800s, a loosely knit group of Montreal merchants formed the North West Company (NWC) and began trading throughout what is now Canada and the north-central United States.[37] The NWC was at a disadvantage in competing for furs with the HBC, whose charter gave it a virtual monopoly in Rupert's Land. But the NWC boldly laid claim to the region after risky overland and sea expeditions from Montreal to James Bay in 1803, and conflict between the two fur empires quickly escalated. Even before these events in 1795, the "Nor'Westers," as they were called, accounted for almost four-fifths of the fur trade in Canada, far surpassing the HBC. At one point during that decade, the NWC also had three times the men.[38]

Despite their ubiquity, NWC's Montreal headquarters put them at a long-term disadvantage, as round trips by canoe to key locations that were relatively close to HBC forts like Lake Winnipeg, Athabasca, and English River could not be completed in one season before the rivers froze. This necessitated the inconvenience of storing trade goods for two years. The logistical problems with transport, combined with overharvesting of furs and various scandals, expedited the NWC's demise. In 1816, some of the wealthiest NWC partners left the company, unsure of its future viability. Conflict between the NWC and HBC led to bloodshed a few years later, and the secretary of state for war and the colonies, Henry Bathurst, ordered the rivals to cease hostilities. When the British government applied pressure with additional regulations governing fur trade in the area, the NWC merged with the HBC in 1821. Decades later, the HBC's vast territory became the largest private landowner in the newly formed Dominion of Canada. It remains North America's oldest multinational resource and trading company and has since diversified, owning and operating retail stores like Hudson's Bay, Home Outfitters, Lord and Taylor, Saks Fifth Avenue, and Zellers.

The depletion of furbearers during English-French competition in North America set the stage for the unsustainable harvest over the next century that eventually necessitated the introduction of nutria from South America to rejuvenate stock. The HBC and NWC had already skinned hundreds of millions of furbearers throughout Canada and the Great Lakes area during the 1700s and 1800s. One report had the HBC selling 14,730 martens; 1,850 wolves; and a whopping 26,750 beaver pelts—all in November 1743 alone. Indeed, the beaver skin was used as a bartering standard and default currency by the HBC.[39] Perhaps Obbard et al. said it best in their seminal study on furbearer harvests:

> Once humans began hunting and trapping for commercial purposes, a significant change occurred: market demand replaced individual need in controlling the size of the harvest. This shift held serious consequences for the survival of several animal species because the forces of the market system, more so than those of individual need, had the potential to cause a major decline in populations or even the extinction of a species.[40]

*Castor canadensis*, the American beaver, was definitely one of these species, as their overharvesting was particularly severe. Popularity of beaver hats as a status symbol continued through the early 1800s, with fashions worn about town including the Paris beau (1815), Wellington (1820–40), D'Orsay (1820), Regent (1825), and Clerical (eighteenth century); other styles like the continental cocked hat (1776), Navy cocked hat (nineteenth century), and the Army shako (1837) indicated military status. With demand for beaver fur at an all-time high, the decline of beaver populations continued precipitously as a new country called the United States of America expanded.

## How the West Was Lost

. . . and that claim is by the right of our manifest destiny to overspread and to possess the whole of the continent which Providence has given us for the development of the great experiment of liberty and federated self-government entrusted to us.

—American columnist and editor John Louis O'Sullivan, addressing an ongoing boundary dispute with Great Britain in the Oregon Country in the *New York Morning News* on December 27, 1845

The arrival of nutria to North America in the late nineteenth century injected stock into a highly competitive market that demanded an omnipresent influx of fur despite a declining supply. Nutria introductions became especially crucial for the fur industry's survival after the voracity of America's pelt trade obliterated beavers along with almost every other furbearing mammal on the continent. Probably the most significant event with regard to fur trapping in the United States was the Louisiana Purchase of 1803. "Louisiana" encompassed much more than the state with that name today; the territory was a vast area west of the Mississippi and Missouri Rivers, stretching from the Gulf of Mexico to the 49th parallel. This historic transaction between the eventual French emperor Napoleon Bonaparte and American president Thomas Jefferson doubled the country's size, opened an immense region for settlement, and made myriad wildlife resources available to American trappers.[41]

The purchase also rejuvenated New Orleans as a valuable shipping port for furs. New Orleans became even more relevant when, in 1764, Saint Louis was founded almost 700 miles to the north as a trading post by fur traders Pierre Laclède and René Auguste Chouteau. The pair intended to "establish one of the finest cities in America," and central to their plan was trading in furs.[42] Four years earlier, the New Orleans firm of Maxent, Laclede and Company secured exclusive rights from France to trade with Native Americans in the Missouri River valley and the territory west of the Mississippi River. The Louisiana Territory, which included modern-day Saint Louis, would change mother countries twice before it became a US territory, but many French traders remained there and sold through Saint Louis rather than traveling to far-flung posts in French Canada. New Orleans therefore served as a crucial station for pelt exports acquired from traders further north.[43]

Meanwhile, American traders became eager to move westward, hoping to exploit an unharvested population of beavers and other furbearers. So westward they went, starting with the 1804 Lewis and Clark expedition, sponsored by President Thomas Jefferson to explore the Louisiana Purchase and to assess its usefulness for settlement and trade. Furriers like Manuel Lisa of Saint Louis followed the expedition, which left "the finest city" to journey up the Missouri River and west to the Pacific coast. The outfit returned two years later with Lisa and his associates, including Meriwether Lewis's son Reuben and René Auguste Chouteau's son Jean and grandson Auguste, who formed the St. Louis Missouri Fur Company (SLMFC) in 1809 to trade in the upper Missouri Valley.[44]

As a result of opening the west to settlement and formation of the SLMFC, tens of thousands of beavers were harvested. The Missouri River fur trade grew, with efforts to cut out middlemen increasingly resulting in white men replacing Native Americans as trappers. Historians tell us that these "mountain men," who eventually traveled into the Rocky Mountains to pursue their trade, were a fast-living, unattached, vainglorious, and free-spirited crowd that often spent their full year's earnings on minimal supplies, whiskey, and a week of wild living before heading back into the mountains for the next year's trapping season.[45] According to essayist Washington Irving:

> There is, perhaps, no class of men on the face of the earth, says Captain Bonneville, who lead a life of more continued exertion, peril, and excitement, and who are more enamored of their occupations, than the free trappers of the West. No toil, no danger, no privation can turn the trapper from his pursuit. His passionate excitement at times resembles a mania. In vain may the most vigilant and cruel savages beset his path, in vain may rocks and precipices and wintry torrents oppose his progress; let but a single track of a beaver meet his eye, and he forgets all dangers and defies all difficulties. At times, he may be seen with his traps on his shoulder, buffeting his way across rapid streams, amidst floating blocks of ice: at other times, he is to be found with his traps swung on his back clambering the most rugged mountains, scaling or descending the most frightful precipices, searching, by routes inaccessible to the horse, and never before trodden by white man, for springs and lakes unknown to his comrades, and where he may meet with his favorite game. Such is the mountaineer, the hardy trapper of the West, and such, as we have slightly sketched it, is the wild, Robin Hood kind of life, with all its strange and motley populace, now existing in full vigor among the Rocky Mountains.[46]

Indeed, the lives and times of these mountain men are the material for countless books and movies. Well-known folks from this era include Jed Smith, Bill Sublette, Jim Bridger, Kit Carson, Joe Walker, and Tom Fitzpatrick.

The free-trapper lifestyle was risky though, as relations with Indians in the Missouri Valley were often testy. A letter from an outpost to one of the SLMFC partners demonstrates the conflict:

Three Forks of the Missouri
April 21, 1810
Mr. Pierre, Chouteau, Esq., Dear Sir and Brother-in-law:
I had hoped to be able to write you more favorably than I am now able to do. The outlook before us was much more flattering ten days ago than it is today. A party of our hunters was defeated by the Blackfeet on the 12th inst. [installment.] There were two men killed, all their beaver stolen, many of their traps lost, and the ammunition of several of them, and also seven of our horses. We set out in pursuit of the Indians but unfortunately could not overtake them. We have recovered forty-four traps and three horses, which we brought back here, and we hope to find a few more traps. . . .

Besides these two, there are missing young Hull who was of the same camp, and Freehearty and his man who were camped about two miles farther up. We have found four traps belonging to these men and the place where they were pursued by the savages [Blackfoot Indians], but we have not yet found the place where they were killed.

In the camp where the first two men were killed we found a Blackfoot who had also been killed, and upon following their trail we saw that another had been dangerously wounded. Both of them, if the wounded man dies, came to their death at the hand of Cheeks, for he alone defended himself.

This unhappy miscarriage causes us a considerable loss, but I do not propose on that account to lose heart. The resources of this country in beaver fur are immense. It is true that we shall accomplish nothing this spring, but I trust that we shall next Autumn. I hope between now and then to see the Snake and Flathead Indians. My plan is to induce them to stay here, if possible, and make war upon the Blackfeet so that we may take some prisoners and send back one with propositions of peace—which I think can easily be secured by leaving traders among them below the Falls of the Missouri. Unless we can have peace with these [men] or unless they can be destroyed, it is idle to think of maintaining an establishment at this point.[47]

Poor relations with Indians kept many trappers away from the intermountain west and confined most trade to the upper Missouri region for the next decade or so. The obvious exception was the Pacific Northwest, which became the platform for the man who was perhaps most responsible for countrywide annihilation of beavers. His name was John Jacob Astor, a

German-born American businessman, merchant, fur trader, and investor who moved to the United States following the revolutionary war. Having emigrated to England as a teenager to work as a musical instrument manufacturer, the ambitious Astor capitalized on the 1794 Jay Treaty between England and the United States, an agreement that opened novel fur markets in Canada and the Great Lakes region. While in London, Astor negotiated a contract with the NWC, importing beaver furs from Montreal to New York and shipping them to Europe. Supplemented with a side venture trading teas and sandalwood with China, Astor's fur business was successful to the tune of nearly $250,000 USD—the equivalent of hundreds of millions of dollars today.[48]

When the US Embargo Act of 1807 closed off Astor's trade with Canada, he persevered. From his residence in New York City, Astor established the American Fur Company (AFC) to operate in the Great Lakes region with the permission of President Thomas Jefferson. Thus, by the time the NWC was merged with the HBC in 1821, Astor had built a monopoly in the Great Lakes fur trade. Astor had also reached what would become the country's western limits in 1811 by founding the first American community on the Pacific Coast, a trading post on the Columbia River at Fort Astoria. Members of the expedition he financed to reach this outpost discovered the South Pass, a route by which settlers on the Oregon, Mormon, and California Trails passed through the Rocky Mountains.[49]

Astor later formed subsidiaries like the Pacific Fur Company (PFC) and the Southwest Fur Company (SFC), dabbled in the smuggling of Turkish opium to China and England, and established the Astor House on strategically located Mackinac Island as a company headquarters. He became the first multimillionaire in the United States and was the wealthiest person in the country at the time of his death in 1848 with an estate worth over $20 million USD—almost 1/100th of the American gross domestic product, and by some estimates equivalent to $130 billion dollars today.[50] Almost as remarkable as Astor's wealth from the fur trade is his decision to get out of the business in 1830 to become a patron of the arts and to build up the land in New York City he bought from Vice President Aaron Burr.[51]

Astor had a keen business sense and emancipated himself at precisely the right time from a fur industry that made him wealthy beyond imagination. His efforts, combined with those of the HBC and NWC, meant that the beaver was for all practical purposes trapped out by the 1830s. Thus, fur trading became increasingly dependent on prey that was growing scarcer. Were it not for the emergence of silk and cholera, the beaver might have been annihilated completely.[52] A rapid decline in beaver hat demand coincided with interest in silk hats, which were made from inexpensive Chinese materials imported to Europe and thought by many to have a more elegant color and feel. When Asiatic cholera reached Europe in 1830, some folks stopped buying furs because they blamed some of its spreading on contaminated foreign clothes. At the beginning of that decade, beaver pelts traded in Saint Louis fetched a high price of six dollars per pound. A few years later, the price plummeted by 50 percent.[53]

The other thing that saved the beaver from extinction was the realization by fur dealers that nutria could be used as a viable substitute for beaver. Furriers recognized that nutria fur was shorter, looser, and more lightweight than beaver fur, especially if sheared or plucked from the animal.[54] But the fur was also warm and fun to wear. Similar in texture and color to beaver, the soft, sporty nutria fur was ideal for vests or coat linings, and was usually dyed black, brown, or beige. Some innovative furriers also tried using unplucked or unsheared nutria, which created a natural look accentuated by thick and glossy guard hairs, a dense underfur, and a light-yellowish or red-brown color. Of course, the prolific reproduction of nutria also made them perfect for exploitation as a perpetual supply of fur.[55]

Because beavers had been hunted indiscriminately for two centuries already, nutria were more plentiful and cheaper than beaver pelts. Thus, nutria exports from South America increased exponentially during the next few decades. With the help of herding dogs, Argentinean natives slaughtered millions of nutria for meat and either wore or exported their skins, which were worth 50 cents to $1.50 apiece, according to the quality of the hide. Argentina and Uruguay exported thousands of nutria skins per year during the nineteenth century, and other countries were quick to catch on to the profitability of nutria-related operations.[56]

By the end of the century, France was an especially salient buyer in the international nutria market. The South American rodent even caught the attention of French academic circles, and *La Société Zoologique d'Acclimatation* imported nutria with the intent to breed them as early as 1885.[57] The reasons for *La Société's* nutria introductions, however, were philosophical as well as entrepreneurial. Those philosophies may have been rooted in the approach of French naturalist Étienne Geoffrey Saint-Hilaire, who suggested a formal name for nutria (*Myopotamus bonariensis*) that never caught on, and perpetuated the now widely discredited "use it or lose it" beliefs of Jean-Baptiste Lamarck, another French naturalist who proposed that organisms can pass on characteristics acquired during their lifetime to offspring. Species transmutation—the conversion or transformation of one species into another—was another notion maintained by some

La Société members that is rejected by the current biological community. Saint-Hilaire's son, Isidore, founded *La Société Zoologique d'Acclimatation* in 1854 and followed in his father's antediluvian footsteps by maintaining that humans and animals could be forced to adapt to new environments.[58]

Alfred Russel Wallace, who independently conceived the theory of evolution through natural selection concurrently with Charles Darwin, defined *acclimatization* in his entry in the eleventh edition of the *Encyclopaedia Britannica* in 1911. Wallace differentiated acclimatization from domestication and naturalization, noting that a domesticated animal could live in environments controlled by humans. He stated that acclimatization involved "gradual adjustment" and suggested that this could be a component of naturalization.[59] But the idea among *La Société* members was associated tightly with the eventually rebuked ideas of Lamarckism. Indeed, Wallace noted that Charles Darwin, whose ideas on evolution withstood time's test and were more in line with his, denied the possibility of forcing individual animals to adjust, although he allowed for the possibility that variations among individuals could provide some individuals with the ability to adapt to new environments.[60]

Either way, *La Société* valued taking animals from one place and putting them in another. Their acquisition of nutria probably contributed to the presence of nutria in western Europe, if not directly through releases, then indirectly by encouraging exchange of animals across international boundaries. For example, *La Société* established a branch in Algeria and established the *Jardin d'Acclimatation* in Paris by 1861 to showcase animals, plants, and even people from other lands. They also awarded medals to members for establishing breeding animals, with the criteria of maintaining at least six specimens with two instances of breeding in captivity.[61] Whether from *La Société*'s stock or not, nutria were bred in France and appeared in the country's wilds around 1882, with fur farming beginning there a few decades later.[62]

Other acclimatization societies in Britain, Australia, and New Zealand may have obtained nutria, but this remains unclear. Nevertheless, it is apparent that some introductions of aesthetically or commercially valuable species that were consistent with the vision of these societies created ecological disasters. Introducing rabbits (family Leporidae) to Australia and New Zealand is an infamous example. Brought to the lands down under by ships and spread with the help of premeditated releases, the rabbits were so fruitful that they necessitated passage of a Nuisance Act in 1876 to legislate their control. After failing to learn the appropriate lesson, weasels were released to control the rabbits, an action described by one scientist as an "attempt to correct a blunder by a crime."[63] And then there was the nostalgia-inspired introduction of five dozen European starlings (*Sturnus vulgaris*) to New York City in 1890 by American Acclimatization Society (AAS) president Eugene Schieffelin, who was leading the society in a peculiar attempt to reintroduce every bird species mentioned in the works of Shakespeare into North America. Resulting from the AAS efforts were perhaps 200 million starlings with almost no natural predators.

It eventually became clear that moving animals from one continent to another was imprudent. Thus, from the 1890s to the 1910s, various naturalists such as T. S. Palmer and the editors of *Avicultural Magazine* wrote in opposition of animal introductions. As a result, the concept of an "ecological balance" where organisms interact with their environment and each other became more prevalent among naturalists. With this change in approach, quarantine regulations were developed and many acclimatization societies morphed into fish and wildlife organizations.[64] Needless to say, folks decades later ignored the known danger of taking animals outside of their natural habitat and cavalierly introduced nutria to Louisiana and other states.

In the meantime, nutria importations continued a cascade of efforts to fulfill the many new markets that would emerge from a changing international fur industry in the face of a declining supply of beaver pelts. One trend was that furriers went to the sea to capture marine mammals because fewer furbearers remained on land, a trend that was especially prevalent in Russia. Until the 1830s, North American beaver flooded the European fur market, with pelts usually imported by England or France to be sold there or exported elsewhere on the continent. Russia was the primary buyer of these western European exports, often obtaining *castor sec* (French for *dry beaver*) pelts. The *castor secs* were never-worn furs that had been scraped clean, and required additional processing to prepare them as felts for hats. A combing technique developed by the Russians helped prepare the *castor sec* pelts by separating the desired beaver wool from the outer guard hairs, and until the combing process was known to western Europeans, the French and English exported *castor sec* pelts to be combed in Russia and reimported the combed pelts. But the Russians also traded furbearing marine animals such as seals and otters out of the Bering Strait. The practice had already started in the 1740s with selling otter pelts to China, but the operation became even more lucrative when the Spanish in California joined in, obtaining otter and seal pelts from local Aleut Indians.[65]

The United States also caught on to the marine fur supply, utilizing ships that traded with China in the late 1700s. American crews wanted to participate in the fur trade, but found that their access to pelts was blocked by the Spanish and Russians until Captain Robert Gray found an entrance to the Columbia River and surveyed the lands into which it drained. By the early 1800s, the Russians made an arrangement with the Americans that involved Aleuts hunting off the coast of California in American ships. The arrangement worked so well that in 1812 the Russians established Fort Ross to serve as a base for the Aleut hunters, therefore eliminating the need for ships.[66] Even Mexico got into the fur trade during this period. Upon their newly gained independence, Mexico permitted hunting by Russians and Americans on the coasts of lower California in return for 50 percent of the profits. Taken together, these machinations all but eliminated several species of seals and otters before California was ceded to the United States in 1848. By the time marine mammal hunting was regulated by international treaty in 1911, it was too little, too late. As for Alaska, the supply of furbearing sea mammals became so depleted that Russia became disinterested in maintaining the territory, later selling it to the United States.[67]

Another post-beaver trend in the international fur trade was the hides of bison (*Bison bison*), commonly called buffalo, which made sturdy leather. Native Americans had a sacred relationship with the bison, using every part of the carcass for their livelihood, and this made bison a focus of anti-Indian sentiment. Once roaming the plains in massive herds that seemed inexhaustible, the story of the buffalo's demise is well documented. In an 1875 speech to the Texas legislature, General Phil Sheridan highlighted this demise when he took the floor in opposition to a bill that would confer protections to the buffalo:

> These men [the buffalo hunters] have done more to settle the vexed Indian question than the entire regular army has done in the last thirty years. They are destroying the Indians' commissary. Send them powder and lead if you will, but for the sake of a lasting peace let them kill, skin and sell until the buffalo are exterminated. Then your prairies can be covered with speckled cattle and the festive cowboy who follows the hunter as the second forerunner of an advanced civilization.[68]

White hunters used the new railroad to penetrate the West and enact Sheridan's philosophy, slaughtering buffalo unabated by the tens of thousands. By the mid-1800s, almost no buffalo remained east of the Mississippi River that La Salle so famously claimed for the French. Wanton killing of buffalo continued as professional hunters trespassed on Indian land and killed over four million bison by 1874. Bison were wiped out for all practical purposes about a decade later, and the so-called Indian problem was "solved" when Native American leaders resigned their people to life on crudely organized reservations.[69]

Biologists say that only humans are capable of affecting the environment more than beavers. These flat-tailed rodents shape the mountain landscape by damming streams and creating huge ponds, providing a place to build their lodge and homes for hundreds of other species. And with the fur trappers and traders that the beaver lured deep into the North American wilderness, probably no other animal has shaped history more either.[70] Nothing short of death will stop a beaver from felling trees and creating a dam when it hears the sound of running water, and a somewhat similar determination drove the animal's hunters from sea to shining sea. Thus, from around 1500 to 1900, tens of millions of beavers were killed across North America; probably around 100,000 were left, most hidden in the Canadian interior, and other furbearers suffered similar losses.[71]

One might think that the combined obliterations of beavers, seals, otters, and buffalo would have resulted in a crippled fur industry. But amazingly, the en masse killing of wild land animals for their fur or hide was hardly finished. The nucleus of the industry simply moved south, with furriers undeterred by furbearer scarcity in the north. Specifically, the fur industry's rejuvenation arose from the Gulf Coast region, which was originally held in Spanish land grants with legal title held by the federal government. But that was before 1850, when the US Congress passed the Swamp Land Act (SLA) and ceded large tracts of real estate to several states, including Louisiana, with the states retaining the right to sell. Following the SLA, Louisiana sold massive parcels of land to various individuals for 10 to 20 cents an acre. Inevitably, some of these newfound landowners could not manage this large amount of land and reneged on their property taxes, causing ownership of the land to revert to the state.[72]

In 1890, the Louisiana legislature granted large sections of land to various levee districts, from which considerable revenue was gained for flood control. With these lands out of private hands, the stage was set. Around 1895, just after wild buffalo were all but obliterated from North America, folks began to refocus on the hidden wealth of the Gulf Coast marshlands. With limited opportunity for trapping and trade in the north, the eyes of the fur industry looked to the Louisiana wetlands—with its alligators, snakes, raccoons, minks, otters, and muskrats—and, eventually, to its nutria.

## Pitchforks and Dogs

In the absence of hearty beaver, seal, and bison populations, Louisiana's muskrats became the heartbeat of America's fur industry. Once muskrats crashed the American fur scene at the turn of the twentieth century, New Orleans also returned to its status as a major shipping port. The Big Easy's central location meant that pelts acquired in the continent's heartland could be shipped to overseas markets without the production of moving them all the way to coastal ports along the Saint Lawrence River or in New York.[73]

Muskrat availability also coincided with a continent-wide scarcity in game, which meant that the southern Louisiana marshland became arguably the number-one fur area of North America. During the early 1900s, Louisiana's fur industry grew to encompass more than 1,000 fur dealers or buyers and over 20,000 trappers.[74] The days of these trappers were a sort of throwback to the days of the mountain men in the west. They might wait for the cold, locate a mound of grass constructed by a muskrat colony, and start setting traps in the nearby trails. Traps would be left overnight because muskrats are nocturnal, and the sunrise would bring success—captured 'rats. It was not unusual for a trapper working in a remote area to set up a camp in the marsh with his family and trap there for months. It wasn't a job—it was a lifestyle, an entire way of life.

When a trapper traveled back to camp, he would skin muskrats in the field and carry the pelts in a sack to reduce his load. A quick-skinning trapper could skin a muskrat in seconds and clear hundreds of traps daily. Good ol' skinnin' knives would cut from the tail to the foot and down the back of each hind leg, and trappers would then reach under the skin on the back to separate it from the carcass. As required by fur buyers, the head was cut so that the ears remained on the pelt. In camp, furs were washed, scraped, and put on a stretcher to dry before storage in a cool, dry place. Negotiation with a buyer usually followed shortly thereafter.

The earliest record of muskrat pelt trading between New Orleans and markets on the upper Missouri occurred in 1879.[75] But muskrats were hardly mentioned in the journals of early settlers to Louisiana, suggesting that human contact with the rodents was probably intermittent until around 1910, when the muskrat trade was established.[76] Whether this is because muskrats became more plentiful during this time is not eminently clear. If muskrat population increases occurred, the reasons likely involved changes in predator abundance and marsh burning by locals.

Alligator hunting and trade in alligator products became popular in the early twentieth century, and this probably allowed muskrat populations to

Skinnin' a 'rat. A trapper sits and skins a 'rat during one of Louisiana's heydays of muskrat trapping in the 1920s. Courtesy of Louisiana State University Special Collections: Louisiana Department of Wildlife and Fisheries Slide Collection.

thrive with fewer predators. Alligator hides were made into leather; alligator tails provided meat that was eaten or rendered into lamp oil; alligator teeth were used in jewelry or ground up to make powder charges for muskets; and alligators musk glands were used to make perfume.[77] Thus, alligators became scarcer, and many scientists maintain that a vicious cycle began when hunters realized that they could find the remaining gators by burning the marshes. Louisiana trappers traditionally hunted the largest and most visible alligators with a method called *shinning*, in which trappers slowly waved a lamp from a boat at night to spot the reflection of alligator eyes on the water's surface. Shinning was an efficient method that was largely responsible for the high level of taken alligators from the 1880s to the 1930s, estimated by one naturalist to be 3–3.5 million individuals.[78]

As alligators became more dispersed across the marsh, hunters sought to make all gator dens more visible on the surface and the marsh easier to walk through by burning excess vegetation. This created the two-pronged effect of fewer alligators to predate muskrats and holding the marsh in an

earlier successional stage of vegetation. The ecological situation was optimal for muskrats because it improved the quality and increased the amount
of Olney's three-square grass (*Schoenoplectus americanus*), a subclimax species that was their go-to forage. A muskrat population explosion followed,
and natural blazes that were triggered by lightning cleared more marshland
and exacerbated the growth in muskrat numbers.[79]

From the southeastern marshes of the Mississippi delta, muskrat populations expanded westward toward the Texas border and began gobbling cattle range. Hunting muskrats therefore took on an enhanced level of social
and economic importance, as ranchers in Cameron Parish were left with no
choice but to implement a five-cent bounty on each muskrat killed. According to one report:

> Cattle range by 1912 was being destroyed by eat outs caused by exploding
> populations [of muskrats] in the southwestern Louisiana marshes. The musk-
> rats were hunted with pitchforks and dogs.[80]

It was not long before the "pitchforks and dogs" were traded for Conibear
traps, as trapping became so widespread that Louisiana started requiring
trapping licenses for an open season on mink, otters, raccoons, and the
increasingly abundant muskrat lasting from November to February. During the 1913–14 season, a whopping 5 million furbearers were trapped in
Louisiana, 4.25 million of which were muskrats.[81]

This trend continued as the New York fur market realized the quality of
pelts available from southern Louisiana. To meet demand, massive trans-
ports of muskrat pelts north following huge muskrat harvests became a way
of life in Louisiana by the 1920s. Trappers cleaned up and prices soared—
outdoorsmen routinely made more than $10,000 during a 70-day trapping
season, with pelts that were once worth only two cents going for $2.50. Over
20,000 trapping licenses were sold, more than 1,000 fur dealers and buyers
practiced their trade, and 6.77 million pelts from muskrats and other furbearers were taken to the tune of $6.49 million—and that was just during
the 1924–25 season.[82] By this time, Louisiana led the continent in fur production, with 90 percent of the take coming from muskrat pelts. For comparison, the entire Dominion of Canada harvested 3.8 million pelts during
1924–25—not even 60 percent of the take in Louisiana.[83]

Ever-increasing involvement in the fur trade inspired the Louisiana
Department of Conservation (LDC) to proudly proclaim, "Today [1931],
Louisiana annually produces more pelts than any other U.S. state or Cana-
dian province, and leads all states in the production of muskrats and mink."[84]

Indeed, trapping during the 1930s and '40s took around four million musk-rat pelts statewide, and by the 1940s, southern Louisiana was home to a $6–10 million fur industry.[85] This prosperity built until its pinnacle in 1945, when over nine million pelts of muskrat generated more than $12 million—more production than the rest of the United States combined.[86]

## Can't Get Enough

After a lull in the late 1800s, the early 1900s was a heyday for Louisiana muskrat traders and the entire fur industry. Fur was *en vogue* worldwide during the 1920s as a display of wealth and style, and following World War I, regular folks could afford furs that used to be available only to the socioeco-nomic elite. A profitable retail fur industry had been established in North America and Europe during the 1800s, inspired by the success of a Parisian merchant company built by entrepreneur Revillon Frères, who recognized the fur-for-the-masses market and established boutiques in France, London, New York, and Montreal. When the automobile became commonplace in the 1900s, fur coats that were long came into style, as they offered pro-tection from the cold air that passed while riding in a car. Raccoon coats became especially fashionable in the 1920s and '30s, and they were worn by men and women alike.

One researcher concluded that two out of three women chosen at ran-dom off the street in any English town would be wearing a fur coat, or at least one trimmed with fur.[87] This statement demonstrated an exorbitant demand, but worldwide desire for fur became even more astronomical when Hollywood stars started wearing mink coats and fox furs. Ankle-length fur coats made of long, vertical strips of fur emerged as the coat of choice for many fashionable, elite, and wealthy women. White or gray fox coats were particularly modish.[88] Lili Damita, for instance, wore fox in the 1932 movie *The Match King*; Mae West and Gertrude Michael donned white fox in *I'm No Angel* the next year. The Oscars were also rife with fur styles during this era; Marie Dressler wore an ermine coat to the ceremony in 1931. Four years later, Claudette Colbert accepted her award from Shirley Temple with a white fur coat draped over her arm.[89]

The fur fashions soon expanded beyond coats, as it became chic to wear animal skin as a scarf or to carry "fur muffs" as an alternative to a purse or handbag. As fur items became trendier, the animals used to satisfy demand became increasingly exotic. Coats, caps, jackets, wraps, and collars were made from myriad animals such as marmots, goats, squirrels, ermines, foxes, raccoons, ocelots, panthers, chinchillas, moles, opossums, otters, bears, seals, rabbits, weasels, and others.[90]

Even when tougher times came with the 1930s, folks nationwide looked to furs as a symbol of wealth that most could not obtain. This remaining positive view of wearing fur allowed marshland trappers an opportunity to make a decent living when many Americans were barely subsisting. Muskrat prices tumbled to well under a dollar during the Great Depression, but with consistent harvests of two to four million furbearers, most of which were muskrat, the coastal economy was able to remain afloat.[91]

No matter which direction the fur markets were moving, Louisiana's fur trade was changing with the times. Coastal landowners wanted a piece of the fur profits, and they began leasing their lands to trappers directly or renting their properties to leasing companies who dealt with the trappers.[92] This practice ensured that the property was worked by licensed trappers, eliminated competition between trappers, and reduced the number of poachers.[93] But it also effectively ended the free-trapping era in Louisiana. A 1943 *Saturday Evening Post* article reminisced about the laissez-faire days of trapping wistfully:

> Dawn was coming. It broke dimly on the trapper when the landowner waked [sic] up and rubbed his eyes at the unbelievable thing that was happening in his watery wastelands. It burst on him with a vivid vengeance when fur dealers of the East descended on Louisiana, discovered both trapper and landowner, and set out to make a deal. The trappers picnic was over. The landowner was coming into his own.
>
> But the sun shone brightest on the middleman who leased the marshlands from the landowners and re-leased it to the trappers. The landowners, eager to find some way of systematizing the industry and gathering their share of the fabulous harvest, willingly signed over their rights to the several syndicates which were formed by Northern capitalists. In a short time the whole coastal region was in the hands of a few big lessees.[94]

By that time, nearly all of southern Louisiana's marshes were controlled and managed by landowners for trapping purposes, and the ways of the independent trapper were gone forever.[95]

The other major change for the fur industry came in response to a decreasing supply of wild furbearing mammals nationwide. Furbearers, after all, had experienced wholesale slaughter over the past three centuries and their populations could no longer meet international demand. So, hoping to rejuvenate wild stocks, entrepreneurs increasingly looked to fur farming, an option that had only recently emerged as viable. The rationale for fur farming was to provide additional stock for trade and possibly wild releases, which could allow trappers to continue their traditional lifestyle. Fur farming would also respond to a demand for furs that was increasingly exotic, one that held the potential to create new and exciting markets that would not be satisfied by Louisiana muskrat alone. A niche developed for materials that resembled beaver and muskrat but were softer, sportier, and less expensive. And the fur industry often looked to nutria to fill it.

Nutria furs were a product with an international market already in place. Because of their popularity as an imported replacement for beaver starting in the 1830s, Americans started to raise nutria in domestic fur farms. Some farms even considered nutria to be an upgrade from muskrat because the larger surface area of nutria pelts compared to muskrats provided manufacturers with more usage options. Nutria pelts were particularly easy to raise and sell in the Sunbelt because the subtropical and humid climate resembled the nutria's native South American home, and because nutria physically resembled the beloved muskrat. And of course, everyone in the fur industry loved that nutria were prolific breeders, a quality certain to provide farmers with a profitable captive operation.

William Franklin Frakes started the first American nutria farm, and probably the first nutria operation outside of South America, in 1899 on a 160-acre ranch at Elizabeth Lake in northern Los Angeles County, California. The venture was not a rogue one, and was encouraged by Stanford University president David Starr Jordan and legendary US Biological Survey scientist C. Hart Merriam.[96] Will Frakes was an appropriate candidate for the extraordinary project, as he was a natural pioneer and a man of great courage and wanderlust. For example, legend has it that Frakes once killed a bear with a revolver after it knocked him over.

Born in 1858, Frakes explored Argentina and attempted to settle there in the 1890s, until an incident in which he killed his two assailants after he was ambushed and wounded.[97] Frakes returned to the United States with several head of nutria stock he bought in Patagonia and brought them to West Lake Park, Los Angeles, thinking the location would provide the best environment for the nutria to propagate. Apparently, the interest of Mr. Frakes in nutria, which he called his "beaver," was not just for fur—Frakes said he had traveled throughout South America and had hunted and eaten almost every animal there possible, but that no meat he ever tasted could compare with nutria. Starting with three females and a very lucky male, Frakes's nutria ate the roots of plant life from the small, shallow lake in the middle of the desert that was hardly known outside of the immediate area.[98]

The introduction of nutria to California was not the only time that Frakes was associated with a cavalier species translocation. In 1904, Frakes and another dabbler in zoology named Will Mudgett captured a few dozen quail and shipped them to Santa Catalina Island, off the coast of Southern California, to form founder coveys as an academic experiment. Shortly after the quail introductions, Frakes returned to California's mainland and stayed in the area around Yermo, (near Barstow), where he learned to capture bighorn sheep. It was dangerous work, and Frakes's crude methods of capture often rendered him injured. Nevertheless, Frakes tried to domesticate the bighorns, and until 1911 he made failed attempts to experimentally breed them with domesticated sheep under a permit from the State Fish and Game Commission.[99]

With Frakes's innovative spirit came behaviors that some folks regarded as peculiar. Records suggest that Frakes showed the same offhand attitude during his sheep endeavor as he did with the nutria he brought to Los Angeles County. According to "a student of [Frakes's] life":

[Frakes] was undoubtedly the first man to capture full grown Big Horn [sic] and adapt them to living in captivity, or rather semi-captivity. Some of the most tamed he would allow to run loose with his domestic sheep. They would come to his call.[100]

In 1908, Frakes sold two of his sheep specimens to the National Zoological Park in Washington, DC, where they died two years later. Frakes, a self-taught taxidermist, mounted the pair's heads and gave them away to an undisclosed recipient.

As for his rodent-based enterprise, Frakes was forced to move his nutria operation shortly after arrival in California to Camp Cady on the Mojave River. The reasons for this shift remain unclear, but in any case, a great flood in 1906 washed the nutria down into Afton Canyon and they became feral. Whether Frakes's makeshift population of nutria in California proliferated after this incident is unknown, but California had a small feral population of nutria by 1940 that was eventually eradicated, so either Frakes's stock multiplied or subsequent importations occurred.[101]

Whatever fate befell Frakes's nutria, it is clear that other folks followed in his footsteps after his 1920 death in Arizona by trying to make quick fortunes on nutria farms. Among these characters was a Canadian gentleman by the name of C. R. Partik, who received a few nutria from a German acquaintance in 1928. Raising nutria with outdoor pens frozen by the frigid temperatures of interior Quebec, Partik's operation was the basis for a pamphlet published by the *Fur Trade Journal of Canada* on raising nutria, which encouraged fortune seekers to enter the fur trade via the backdoor of nutria farming. Once word spread that nutria were rapid and successful breeders, it was not long before hundreds of nutria fur farms popped up all over North America and abroad. By the mid-1930s, the states of Washington, Oregon, Michigan, New Mexico, Utah, Ohio—and of course Louisiana—were graced with captive nutria.[102]

For a few decades in the early twentieth century, nutria produced as advertised, especially in the United States. But by the mid-1900s, once-plentiful muskrat populations in Louisiana plummeted, along with the national demand for all types of furs. When instant financial gain did not materialize due to fluctuating market dynamics, some farmers quit and emancipated their nutria; other nutria escaped after hurricanes toppled fences. And before anyone knew it, free-ranging nutria were eating crops and wetlands, especially along the Gulf Coast. The space between a few nutria on captive fur farms and thousands of nutria roaming the Louisiana swamps is filled with some fascinating stories of nutria gone wild.

## Chapter 3

# Nutria Gone Wild

I am firmly convinced that [nutria] will, in time, equal in value the muskrat production in this state.
—**Armand P. Daspit,** director of the Louisiana Department of Conservation's Fur and Wild Life Division (1945)

[I] liberated probably one hundred and fifty pairs of these animals in Iberia Parish since 1940, and they have spread to the northern limits of Louisiana and the extreme western limits, and have crossed over Vermillion Bay to Marsh Island. . . .

You know my activity . . . liberating them for the purpose of establishing an addition to our fur industry.
—**Tabasco sauce manufacturer president Edward Avery McIlhenny,** in letters to the *New Orleans Times-Picayune* (1945) and the Louisiana Department of Conservation (1940)[1]

*Avery Island, a 2,200-acre rural area about 120 miles from New Orleans in Iberia Parish, is actually not an island. But it seems like one because the vicinity is encircled by wetlands and is conspicuously higher than its surroundings. Just three miles from the Gulf Coast and deep in Cajun country, the "island" lies on a dome of solid rock salt, formed when ancient seabeds evaporated. With the largest of five salt domes along the Louisiana coast and a unique type of pepper, McIlhenny Company makes its flagship product—Tabasco sauce— for sale in more than 185 countries and territories, with instantly recognizable diamond-shaped labels that are translated into 22 languages and dialects.*

*On its way to becoming a household name, the McIlhennys' hot concoction has been the spice of life for some lofty company. For example, Tabasco sauce came along when England's Lord Kitchener invaded Khartoum in 1898, just a few years after Tabasco was first exported. It was also consumed during the discovery of King Tut's tomb. Tabasco sauce has even been used by astronauts to spice up their food packets in outer space.*

*Travelers to Avery Island can enjoy a factory tour where they learn how Tabasco sauce is made, a gift shop with spicy gifts, and a restaurant with peppery treats. They can tour the gardens, learn McIlhenny Company's history, and sample different flavors of the sauce on crackers or mini-pretzels. But I'm not here for the sauce. I'm here to talk about nutria.*

*On a humid Thursday afternoon in October 2015, I arrive at the only place in the world where Tabasco sauce is made, having driven two and a half hours from suburban New Orleans on Route 90 across the bottom of Louisiana. I continue about a mile down a restricted-access road in a light-blue subcompact economy rental car, making several U-turns and entering the wrong building to ask directions of an office worker before finally coming to a sign marked "Archives." After a right turn into the gravel parking lot, I grab my computer and camera, knock on the door, and enter. I'm ready to learn about the most famous story of nutria gone wild.*

✦ ✦ ✦

Shane K. Bernard knows more about the history of Avery Island than just about any person alive. As historian and curator of the McIlhenny Company's historical documents, Bernard spends his workdays surrounded by Tabasco-related artifacts and oversees the archiving of material owned by McIlhenny Company.[2] Many of the materials have been found in warehouses and the attics of abandoned houses on Avery Island, and some family members and company employees have donated items. The collection of books, papers, and mementos is impressive, with the archives filling several rooms of a small house.

"You wouldn't believe the stuff that just comes up around here," Bernard says intensely while showing me a crumbling manuscript written in cursive script in a room filled with records. "This is the first mention of 'peper [sic] sauce' in the McIlhenny journals."[3] I look intently at the tattered pages in the ripped binding, eyebrows furrowed and head absently moving up and down, unsure of how to respond.

"And here's his original from 1868 right here," says Bernard, holding what appears to be a broken-in recipe. "He [Edmund McIlhenny, the company's founder] just kind of started making the sauce and his family encouraged him because it was tasty. He really didn't make that much money off the idea at first, and how it was made ... it wasn't as scientific back then." I squint my face and nod again. *Tabasco,* I'm thinking. *So much Tabasco....*

"Do you use Tabasco sauce yourself?" I ask, pointing to twin bottles of the elixir on his desk, one red and one green.

"Oh sure, all the time," says Bernard, "on a lot of things. Sometimes I even put it into my banana bread to give it 'kick.' I said when I got this job, that it's so appropriate for me that I would work for this company."

"Yeah, I use it too," I respond, trying in vain to match Bernard's unbridled passion for Tabasco sauce. "Sometimes I put it on corn flakes, or in coffee." It's a lame attempt at a joke, but Bernard, riding his train of thought and entrenched in locating another Tabasco-related document, either doesn't get it or doesn't hear me—probably the latter, unless he takes the comment seriously.

The son of swamp pop musician Rod Bernard, Shane is not just an expert on Tabasco sauce. With a doctorate from Texas A&M, he is a dedicated Gulf Coast local, a learned historian of Cajun and Creole culture, and author of five books with the University Press of Mississippi, including *Swamp Pop: Cajun and Creole Rhythm and Blues* (1996), *The Cajuns: Americanization of a People* (2003), *Tabasco: An Illustrated History* (2007), *Cajuns and Their Acadian Ancestors* (2008), and *Teche: A History of Louisiana's Most Famous Bayou* (2016). Some of his most popular work, however, has come on the subject of the nutria invasion. Before he began working for McIlhenny Company in the mid1990s, Bernard heard the legend about nutria on Avery Island that was well accepted by most of the public. The story blamed Iberia Parish native Edward Avery McIlhenny (1872–1949)—also known as *M'sieu Ned* by those close to him—for the Gulf Coast's nutria invasion, and sometimes even for the introduction of nutria to North America. The story had a "wow" factor because of Ned's influence in the community. Among other accomplishments, Ned McIlhenny presided over the manufacturing of Tabasco sauce at Avery Island from 1898 until his death in 1949, and his father, Edmund McIlhenny, was the sauce's inventor.

For decades, the story of the nutria released on Avery Island read something like this:

E. A. McIlhenny [Ned], the man who ran the Tabasco Sauce business, began to raise nutria on Avery Island as a side venture. He imported a number of nutria from Argentina where they're indigenous, brought them to Louisiana, and raised them in a pen somewhere on Avery Island. A hurricane came along in the early 1940s, knocked down the pen, and the nutria got out. The freed nutria then populated the coastal salt marshes of Louisiana. It was all his fault.[4]

Depending on the storyteller, Ned McIlhenny negligently allowed the rodents to escape during a hurricane, or released them into the swamps maliciously, or both. Bernard saw the legend mentioned in top media outlets like *National Geographic*, the *Washington Post*, and *Audubon*, and in regional newspapers such as the *Baton Rouge Advocate* and *New Orleans Times-Picayune*. Indeed, the story seemed to be readily accepted.

Nutria used to be the most profitable furbearing animals in Louisiana, but they are now universally reviled for gobbling wetlands. To be credited with farming and releasing nutria might have been honorable in the 1930s and '40s, but almost no one today would publicly embrace the notion that they introduced nutria to the wild. Thus, any potential half-truths or omissions of fact needed to be examined carefully before accepting the popular version of events and placing undue blame on Ned McIlhenny, or anyone else. In addition, Ned McIlhenny's involvement with nutria farming needed to be considered within the context of its era, a time when almost no one thought that the fur industry would steeply decline. So, when Bernard stumbled across some "nutria" folders in the McIlhenny Company archives, he decided to put his skills as a research historian to work and either confirm or refute the story. And what he found was a tale that is much more complex than the popular version of events, and a new opportunity to examine how nutria arrived in Louisiana. It was time to do some myth-busting.

Was Ned McIlhenny the first to introduce nutria into Louisiana's swamps? Did he release nutria on purpose? Is M'sieu Ned solely to blame for Dixie's nutria invasion, or did other folks also contribute? What circumstances led the otherwise successful Tabasco mogul to take on a side venture involving nutria, and why introduce nutria on Avery Island? If McIlhenny or others introduced nutria to Louisiana, why did they do it? And, perhaps most important—if Ned McIlhenny was not the first to introduce nutria to Louisiana, then why was he charged with the blame? The answers to these questions are complex—and sometimes as wild as the propagating nutria.

### Saucy Tale

One reason that the previously standard account of Louisiana's nutria introduction became popular is the spotlight on the McIlhenny family and their company. Folks like the sauce, but the Tabasco operation's allure goes deeper than tasty peppers. The McIlhenny Company's founding is a quintessential American tale of hard work and building a business from the ground up. For loyal Southerners, the company is also entrenched in the region's recovery from the Civil War era. Thus, telling about a McIlhenny family member who introduced nutria to Louisiana is a saucy tale—even if the tale is not quite accurate.

Perhaps no Iberia Parish resident has been more influential than Edmund McIlhenny. Born in 1815 to a middle-class family from Hagerstown, Maryland, he moved to New Orleans around 1840 for unrecorded reasons and found work in the Louisiana banking industry. With his entrepreneurial nature and keen negotiation skills, Edmund worked his way up the corporate ladder from a bookkeeper position, building a small fortune of around $112,000 (about $2.5 million today) and becoming an independent bank owner. His connections took him to the Bank of Louisiana branch in Baton Rouge where he met jurist Daniel Dudley (D. D.) Avery, sometimes known respectfully by his professional title "Judge," whose plantation on the land called "Petite Anse" would eventually become the home of Edmund's spicy brainchild.[5]

Before D. D. Avery's planation, Europeans encountered Avery Island as early as 1779, when French explorers employed by Louisiana's Spanish administrators described a "so-called isle" near Bayou Petite Anse (meaning *little cove* in Louisiana French). According to Bernard's research:

> The island itself soon became known as Île Petite Anse or Petite Anse Island, though for a short time around 1800 it bore the names Isla Cuarin and Côte de Coiron. The latter two names alluded to the island's earliest known owner, a Dr. Cuarin or Coiron, who claimed the island in the late 1700s. By the early 1800s, however, several pioneer settlers owned parts of the island. These settlers included Elizabeth Hayes, Jacques Fontenette, Alexandre Devince Bienvenu, Boyd Smith, and Jesse McCall.[6]

In 1818, New Jersey native John Craig Marsh purchased a share of the island and began to plant sugarcane there.[7] More than three decades later, Marsh sold the plantation to his son, George Marsh, and his two sons-in-law, Ashbel Burnham Henshaw and D. D. Avery, the latter of whom came into possession of the other's shares.[8]

Sometime after Edmund McIlhenny and D. D. Avery became friends, Edmund named his yacht *The Secret*. As it turned out, Edmund actually was harboring a "secret" that was no longer one after he revealed it in a letter to Avery when his daughter Mary Eliza turned 20 in 1858:

> My long and intimate [social] intercourse with your family has resulted in an honest and devoted love for your daughter, Mary. Save by inference from my

attention, she is unaware of my feelings. I respectfully ask your permission to make them known to her.[9]

Avery initially left the request unanswered because he disapproved of a prospective son-in-law only five years his junior, but finally consented when Mary Eliza threatened to elope. With Avery's blessing, Edward and Mary Eliza were married in 1859 and they settled in New Orleans.[10]

The coming years brought the couple eight children, although two died as infants. Meanwhile, the natural resources of Petite Anse (which was eventually renamed Avery Island) became a flashpoint for Civil War conflict. During wartime, the island's value revolved around the discovery of solid rock salt, eventually a crucial component of McIlhenny's Tabasco sauce, just 16 feet beneath the ground's surface. Confederate officials, operating with their salt supply cut off by a Union blockade, were pleased when Judge Avery told them about the incredible supply on the island. The salt was useful as a preservative and for medicinal purposes, prompting Confederates to call Avery Island a "gift from heaven."[11] Indeed, this was the first sighting of rock salt anywhere in North America.

With the banks inoperable because of the war, Edmund McIlhenny went to work on Avery Island helping his in-laws extract the salt. But Union forces considered Petite Anse a target because of the salt operation. On April 15, 1863, the McIlhennys and Averys were forced to abandon the island for Texas as Union troops approached. Two days later, the salt works were captured for the North.[12]

In Texas, Edmund McIlhenny served as a civilian employee of the Confederate army, first as a clerk in a commissary office, then as a financial agent for the paymaster. When the war ended, he returned to Avery Island with his family and found that the plantation's salt mines and sugarcane fields survived more or less intact.[13] Thus, Edmund McIlhenny tinkered with utilization of the island's natural resources. According to oral tradition, he moved into the Avery residence, tended to the family's fruit and vegetable garden, and began to experiment with growing peppers.[14] When he used them to make a red pepper sauce that the family enjoyed, Edmund began selling his sauce in regional outlets. By making "peper [sic] sauce," McIlhenny made economic use of the island's bounty because rock salt from the mines was a key ingredient in the sauce's formula.[15]

Contrary to popular belief, Edmund McIlhenny made a lot more money as a banker than as a Tabasco sauce manufacturer.[16] He seems to have not recorded a personal account of his Tabasco-related exploits, and even his wife and children disagreed about basic details of the story, such as when and where Edmund obtained the peppers. Tabasco sauce evolved its popularity over time rather than being immediately successful upon placement on the market, so the accomplishment of creating it was not mentioned in Edmund's autobiographical sketch or obituaries when he died in 1890.[17]

Either way, using salt to make Tabasco sauce began a pattern of commercial resource utilization on Avery Island. During the next few decades, Avery Island saw sand, gravel, lumber, and oil production, along with canning of shrimp, oysters, fruits, and vegetables. Sugarcane production occurred until it ceased to be commercially viable around 1925. And, as with most of Louisiana, fur trapping in the nearby salt marsh played a vital role in Avery Island's economy for many years.[18]

Trapping came to the island in part because two of Edmund and Mary's environmentally inclined sons, first John and then Edward (Ned), controlled the Tabasco operations after their father's death. In addition to fighting trademark infringements and tweaking the iconic Tabasco diamond logo to create the version seen today, Ned enjoyed his role as steward of the natural resources, including furbearing mammals, on and around Avery Island.[19] Indeed, Ned was an adventurous soul who enjoyed taking calculated risks, and manufacturing Tabasco sauce was just one outlet for his boundless entrepreneurial and environmental interests. But after forming a successful hunting and trapping outfit on the island, Ned got into trouble with an offbeat side project that did not go as planned—his nutria pens on Avery Island's northern edge.

## Ned's Nutria

So why was Ned McIlhenny originally blamed for the nutria invasion? One reason is Ned's larger-than-life persona and legendary feats as a naturalist. Folks found it easy to associate the expert outdoorsman and his cosmopolitan experiences with the introduction of nutria to Louisiana.

Charismatic and beloved, M'sieu Ned lived life to the fullest with a down-to-earth, spontaneous, self-confident, and unpretentious manner that sometimes bordered on impulsive. He was probably the best swimmer, runner, and hunter in the McIlhenny family. As a vigorous outdoorsman who learned the trade from his father as a boy, Ned often wandered Avery Island introspectively, watching birds, collecting insects, and taking detailed notes of his field observations.[20] Never a man to shy away from risk, Ned also wrestled alligators for the thrill of it. He may have once killed an alligator that measured over 19 feet in length, said to be the longest American gator ever recorded.[21]

M'sieu Ned. A portrait of Edward A. McIlhenny. Courtesy of E. A. McIlhenny Enterprises, Inc., Avery Island, Louisiana.

Larger than life. A display case with Ned McIlhenny's mementos at the McIlhenny Company archives, including a picture of him on an arctic expedition and some books that he wrote about his experiences as a naturalist. Photo by Theodore G. Manno.

Tabasco king. Edward A. McIlhenny (also known as M'sieu Ned, *right*) and Major James B. Pond at Avery Island, March 27, 1935. Courtesy of University of Louisiana Lafayette Special Collections.

As a young man, Ned left classes at Lehigh University to look for more adventure, and found it when he accepted an ornithologist position on an arctic expedition. Bringing supplies and aid to explorer Robert Peary's Greenland ice cap crew onboard the freight steamer *Miranda* in 1894, Ned began learning about the routes taken by several migratory birds, some of which wintered in Louisiana. This accomplishment mitigated the demise of the *Miranda*, which hit an iceberg off the coast of Labrador, headed back to sea after repairs, but promptly struck a reef and sank. All aboard were saved, including Ned, and the ill-fated journey north never infringed on his interest in arctic biology. Three years later, Ned financed his own expedition to Alaska and returned home with almost 1,600 Eskimo artifacts.[22]

Ned believed in the mantra "nature in balance," fishing and hunting sustainably while supporting pro-environment and conservation initiatives. For instance, Ned banded more than 285,000 birds on Avery Island to learn about their social structure, lifespan, and migration patterns, producing more data on birds in the Gulf Coast region than perhaps any other person in his era. Between arctic voyages, he ambitiously tackled saving the snowy egret (*Egretta thula*) from extinction. Ned observed the egrets migrating

Bird City. Written on the original photo from the 1970s is a description of Ned McIlhenny's Bird City: "The Jungle Gardens of Louisiana's Avery Island came alive with a flush of green during the warm, semitropical springtime along Louisiana's Gulf Coast. Reigning over these 200 landscaped acres of gardens is a 10th century statue of Buddha, housed in the pagoda, which is nestled among mossdripping live oaks and giant elephant ear plants. Avery Island, a bird sanctuary for hundreds of thousands of egrets, herons, ibis and others, is also a salt dome and plantation for the peppers which make the world famous taste delight, Tabasco. Avery Island rises from the south Louisiana marshland eight miles southwest of New Iberia." Courtesy of State Library of Louisiana.

to and from Avery Island as a child, but during his lifetime the populations declined precipitously because of ruthless plume hunters, who killed the exotic bird en masse to acquire long feathers for women's hats. The problem was particularly salient in the Gulf Coast region, where poaching was hard to police in thick brush and near inaccessible waterways with overgrown vegetation. Although a hunter himself, Ned abhorred the practices that caused the snowy egret's plight, writing that "this great reduction [was] not due to natural causes, but to the persecution of [the egrets] by man, who has killed them for both sport and profit."[23] He blamed, among others, those who

Buddha. The Buddha statue given to Ned McIlhenny overlooks a pond on Avery Island. Photo by Theodore G. Manno.

created a market for the feathers, chastising the "barbaric love of adornment, which 1,800 years of Christian civilization has failed to eradicate."[24]

While lobbying Congress for measures to protect them as an endangered species, Ned captured eight snowy egrets and created a private wildfowl refuge for them on Avery Island. He cared for them until winter, then freed the egrets to migrate on their usual route to South America. From his bird-banding studies, Ned suspected the birds would return to the same spot on Avery Island every year, and they did. Each spring, the egrets returned to Avery Island in larger and larger numbers.[25] By 1911, the refuge Ned called "Bird City" harbored over 100,000 egrets and was considered revolutionary because of its roots in individual initiative; conservation-oriented President Theodore Roosevelt called Bird City "the most noteworthy reserve in the country."[26]

Avery Island residents also revered Ned because of his exotic travels. The most salient demonstration of Ned's multicultural interests is the six-foot-tall, centuries-old Buddha statue that sits in a glass house overlooking a lagoon near Bird City. According to oral legend, the statue's original home

was the Shonfa Temple, once a prominent shrine in Beijing, and it was supposed to be shipped to New York by order of a general who looted it during
the fall of the Chinese Empire in the early 1900s. But the deal was left uncon-
summated after the general was captured and beheaded, and the Buddha sat
in a warehouse for years. When the warehouse owner discovered the Buddha
and put it up for auction, some of Ned's friends on the East Coast bought it
for him, knowing that their eccentric friend would cherish the gift.[27]

Ned published several books about his studies, including *Bird City* (1934),
*The Alligator's Life History* (1935), and *The Autobiography of an Egret* (1940).
Ned's interests even expanded into African American spiritual music, resulting in his 1933 book titled *Befo' de War Spirituals*.[28] But the same overde-
veloped sense of persistence that accounted for Ned's successes sometimes
drove him to take unnecessary risks—even before he started a nutria farm.
Examples include his injudicious foray into placing products other than his
father's pepper sauce under the Tabasco brand, such as whole pickled peppers, powdered peppers, canned goods, and extracts of plants like vanilla,
lemon, and cinnamon. The side products were unprofitable and resulted in
a substantial amount of debt, which forced Ned to refinance the company
through a well-endowed Philadelphia in-law before refocusing sales efforts
on Tabasco sauce exclusively.[29]

A proclivity for novel or offbeat approaches to business probably led Ned
into his nutria farming side project, and propagating nutria also fell under
his sincere interest in exotic fauna and flora. Indeed, when Ned received his
Buddha statue, he placed it among his 170-acre "Jungle Gardens" where it
remains today among a cornucopia of nonindigenous plants and trees that
he acquired through talking to folks all over the world, some of which are
hybridized results of Ned's breeding experiments.[30] But it was Ned's desire
to be "larger than life" that led him to spuriously accept credit for not just
operating a nutria farm, but also for being the first person to introduce nutria
to Louisiana. Ned's propensity for telling a tall tale was well known among
his family and Avery Island residents. According to Bernard's research:

> Although an accomplished businessman, biologist, and folklorist, [E. A.] McIl-
> henny made no pretense about being a historian. He often embellished stories,
> particularly about himself, in the jovial manner of a seasoned raconteur.[31]

This tendency was abetted by folks who were drawn to Ned's stories of
intrepid exploration, and he was usually happy to oblige them. In perhaps
the strangest example of Ned's aggrandized storytelling, he spoke of helping to rescue over 100 men who were shipwrecked during his arctic voyage,

and added that "one of the rescued men was Jack London, the writer. He was using the name Jack Edwards then. But he spoke with such a Cockney accent, I called him London. He kept the name."[32]

While it is true that Ned McIlhenny helped to rescue a large number of whaling fleet sailors, the details of Jack London's life, including his birth in Oakland, California, do not bear out the other part of the story.[33] But for years the tale stuck with some folks, bolstering Ned's celebrity. "He liked to take a good story and make it better," Bernard tells me, as we examine the location where Ned's nutria pens used to lie on a topographical wall map of Avery Island, "and his experiences probably made it easy for people to believe his stories."[34]

Ned publicly embraced the notion that he was a pioneer in the importation and release of nutria into the wild. But why take credit for such a thing? The answer must be considered within the context of 1930s Louisiana. In a time of unbridled enthusiasm for a fur industry that was left largely unfulfilled because of declining stock, introducing nutria to liven up the trappers would have seemed like a tremendous contribution to society. This reason is probably why Ned bragged about his "accomplishment" to Houma civic leader Eugune Dumez in a 1945 letter that subsequently appeared in the *Times-Picayune*:

> I originally brought 15 pairs of the animals from the Argentine ... [and] have liberated probably 150 pairs of these animals in Iberia Parish since 1940, and they have spread to the northern limits of Louisiana and the extreme western limits, and have crossed over Vermilion Bay to Marsh Island.[35]

Even in this small excerpt, Ned exaggerated in prototypical fashion, reporting five more pairs of nutria than are officially recorded, and misstating that he received the nutria from South America and not from a farm in Louisiana, which was later found to be the actual origin of the stock. Through this letter and other proclamations, Ned seemed to eagerly take credit for importing nutria to Louisiana and then spreading them through the countryside. But Bernard's research into family records reveals some key differences in fact from Ned's accounts.

In his article titled "M'sieu Ned's Rat? Reconsidering the Origin of Nutria in Louisiana," Bernard discusses a pivotal conversation about nutria in 1930 between Ned and Armand P. Daspit, director of the Louisiana Department of Conservation's Fur and Wildlife Division (LDCFW), which indicates that Ned's interest in nutria began a bit more tangentially than he stated in his letter to the *Times-Picayune*. Daspit received a bulletin about nutria from

Buenos Aires, Argentina, and consulted Ned, who was widely known as an expert on furbearing mammals because he hunted skillfully and allowed trappers to harvest muskrat, mink, otter, opossum, raccoon, and skunk on his property. Bernard presents a letter dated October 16, 1930, in which Daspit told McIlhenny, "I am very much interested in introducing nutria into this country and think that if we could get several pairs and put them on our hunting grounds at the mouth of the Mississippi River that they would do well."[36] After having reviewed information that Daspit gathered from Paul Redington, chief of the Bureau of Biological Survey (BBS) in Washington, DC, Ned explained to the director that "sometime ago" he had gathered "quite a lot of data on this subject," and that

> while I believe the animal will do very well if placed on a range that it cannot wander from, such as Marsh Island, I don't believe that it could be propagated properly elsewhere unless enclosed by a fence, as they are great wanderers. They live on the same sort of food that muskrats thrive on, but the cost per pair is very high. . . . I believe should you care to import some of these animals, it would be best to have them enclosed in an area where there is plenty of natural feed, until they develop sufficiently to put them on open range.[37]

Ned therefore seemed open-minded to the introduction of nutria, but his enthusiasm was tempered because of potential costs and their tendency to wander, unlike muskrats, who were generally more philopatric (likely to stay in or return to a particular area). Daspit was interested too, so he wrote his own letter to the BBS in Washington, DC, later that year:

> I have received a bulletin from Mr. G. Muller of Buenos Aires [Argentina] on the nutria and am very much interested in same. . . . I am wondering if it could be introduced in the state of Louisiana. . . . As you know[,] the state owns several wild life refug es [sic] and I feel these would be very good places for trying out the nutria.[38]

The BBS wrote back:

> It may be highly objectionable to turn them loose. . . . Numerous examples exist in this and foreign countries of the introduction of species from one part of the world to another with very disastrous results.[39]

The conversations made quite an impact on Daspit, who 17 years later recounted Ned's words in an article for *Louisiana Game, Fur, and Fish* that possibly added to the legend surrounding Ned and the nutria invasion. Exaggerating Ned's enthusiasm for nutria farming, Daspit said that Ned "promptly procured several pairs and experimented with them on Avery Island. The colony soon grew very large and were liberated in the marshes."[40] Daspit was mistaken—according to Bernard's findings, it would actually be eight years after their exchange of letters before Ned purchased his nutria.[41]

In any case, records indicate that Ned may have conducted independent research concerning the viability for farming nutria as early as 1926 while concurrently warning colleagues of the potential for nutria to become pests. A letter from New Orleans lawyer William Grant to John Dymond, president of the Delta Duck Club (DDC), discussed the possibility of importing nutria from South America for a whopping $150 per pair. Proposing that the club could expect nutria pelts to fetch a price of "$3.50 to $7," and that, potentially, the club could sell excess "breeding stock to other land owners," Grant touted nutria as environmental control, which was a common reason to peddle nutria at the time. Specifically, Grant stated that "if 5 or 6 pairs could be obtained . . . they would help thin out the alligator grass and restore to fur production sections now worthless and barren."[42] Dymond contacted Stanley Arthur, then director of the LDCFW, and Arthur responded that "personally, I am very much interested in this animal," although he added that $150 was perhaps a bit much for an asking price. Nevertheless, Arthur seemed enthusiastic about the prospect of importing nutria, saying, "I strongly advise to make the experiment. Proper permit will be given you from this office and I doubt if the U.S. Department of Agriculture would or could interfere."[43]

Among the Dymond-Arthur letters in the E. A. McIlhenny Collection is an unsigned carbon copy of a letter dated October 29, 1926. Because the collection contains a complete series of communications and Ned's environmental expertise probably placed him in the loop on such matters, Ned probably wrote the letter. "I believe that conditions in Louisiana are ideal for raising Coypu. They are much more valuable than muskrats, are prolific, and if once started will increase rapidly," explains the author, who suggests "Hagenback, the great wild animal man from Amsterdam, Holland," as a potential source from which nutria might be obtained. But the letter ends cautiously, stating, "Several years ago, I wrote the Bureau of Animal Industry, Biological Survey Department [BBS] at Washington and they replied that they were studying the advisability of bringing Coypu to [the United States]."[44]

Thus, Ned was hardly the first person to think of introducing nutria to Louisiana—nor was he, as it turns out, the first person to actually introduce them. It is also possible that Ned's enthusiasm for raising nutria was tempered for a dozen years before he established his Avery Island nutria farm in 1938. So if not Ned McIlhenny, then who was the first person to introduce nutria to Louisiana?

Apparently it was not Grant or Dymond either. Both men appreciated the information from the letter author (probably Ned McIlhenny), and contacted the BBS for additional information about nutria. Acting bureau chief W. C. Henderson replied to Grant in a November 17, 1926, letter with some general warnings about introducing species, suggesting that if he must introduce nutria, then Grant should "confine the coypu rats to a large enclosure or liberate them in a limited area." Grant was then advised of a bureaucratic "15 percent ad valorem duty on the importation of these animals into the United States."[45] Dymond received a reply from bureau chief E. W. Nelson on December 14, 1926, that was even more tentative about the prospect of nutria importation, citing the damage done by "the rabbit in Australia" and "the muskrat in Eastern Europe." And Nelson eerily—even prophetically—stated that he was "inclined to believe that if these fur bearers [sic] are liberated there for experimental purposes they might become pests."[46]

No evidence exists to indicate that the sets of letters from 1926 or 1930 resulted in actual importation of nutria. If Ned was author of the unsigned letter, then he may have, contrary to popular belief, indirectly delayed nutria introduction to Louisiana by providing the Department of Conservation and DDC with the names of officials who were still studying the feasibility of raising nutria. In any case, it seems that Grant and Dymond dropped their pursuit of nutria introduction because of the risks described by state officials in 1926. They never obtained any nutria, from South America or elsewhere—but other folks did.

### Letting the Rat out of the Bag

Documentary evidence indicates that while "M'sieu Ned" was exchanging letters regarding the prospect of bringing nutria onto his property, at least two other private nutria farms began operation in Louisiana. Susan Brote and her husband, Henry Conrad Brote, operated one of these nutria farms in Saint Tammany Parish.[47]

Born on March 2, 1894, in Worcester, Massachusetts, H. Conrad Brote began his seafaring career in July 1914 by enlisting in the United States Navy. He served on various ships, including the battleship *Wyoming*, until late 1918. Following his discharge from the navy, Brote worked as a merchant marine deck officer, first on vessels of the US Shipping Board and later the

French American Line until June 1921. A year later, in June 1922, he entered the employ of the Mississippi Shipping Company; he would remain an employee of that firm, first as a deck officer and later as a master, for the remainder of his seafaring career. Brote died in December 1984 in Saint Tammany Parish.

Brote's nearly 50-year career at sea took him all over the world, but most official documents pertaining to his voyages, now housed in the Earl K. Long Library at the University of New Orleans, describe his experiences traveling between New Orleans and ports in the West Indies or on the east coast of South America. In fact, when Mr. Brote served as a captain in the merchant marine, he routinely traveled between New Orleans and Buenos Aires, near the geographic origin of nutria. A connection between Brote and nutria was revealed when *Times-Picayune* reporter Martha Carr observed that "Brote was a merchant marine officer who imported 18 nutria from South America in 1933, according to his personal cargo logs."[48]

Indeed, my own research confirms that "Chief Mate" Brote indicates on cargo logs with an unknown point of origin from the ninth voyage of the SS *Del Norte* dated June 8–August 30, 1933, that in addition to thousands of bags containing coffee and meat there were "3 Cages 18 Live Nutria" listed "On Deck" as the ship headed northbound to New Orleans.[49] Then, in a later entry from the twelfth voyage of the SS *Del Norte*, which sailed from February 12 to May 9, 1934, Brote submitted a formal change of address to the Bureau of Navigation, moving from 1019 Fern Street in New Orleans to PO Box 162 in Abita Springs, Louisiana.[50] Both records are consistent with Carr's report, and so is a 1945 letter that Bernard discovered from Susan Brote to Ned McIlhenny:

> My husband . . . and I started a nutria farm in Abita Springs in 1933 and had the animals for four years. We raised them very successfully in brick pens but could find no satisfactory market for them. We sold some to fur dealers and other individuals and turned the rest out.[51]

Adding the stated four years to 1933 puts the release in 1937, at least several months before Ned McIlhenny purchased his first nutria in 1938.

But unlike Brote, who traveled to South America and brought the rodents to New Orleans by ship, Ned had little marine experience and apparently never traveled south of the equator. He told Eugune Dumez in Houma that he bought "15 pairs of [nutria] from the Argentine"—but how?

Once again, it seems that Ned was just taking a good story and making it better. Bernard's research into the McIlhenny files indicates that it was another private nutria farm in Louisiana, which was operating in Saint Bernard Parish, from which Ned obtained his first nutria in 1938. The count of nutria was 14 adults and 6 kits for a total of 20—not the "15 pairs" stated in the *Times-Picayune*—and all but two of the nutria were born in the United States. Unfortunately, little is known about this nutria farm, and the owner's identity remains unclear. But papers in the McIlhenny archives reveal the name "A[be] Bernstein," a New Orleans raw fur and wool dealer, who apparently served as an intermediary in the transaction.[52] Advising McIlhenny in early March 1938, Bernstein writes that:

> I am able to purchase the nutria for you for the sum of $100.00 [$5 per animal] and now you will kindly let me know what day you can send your truck to pick them up, as I must let the party know, in order that he may have them ready for shipment.[53]

Several conclusions are in order. First, considering William Frakes's 1899 nutria farm in Southern California, Ned McIlhenny was not even close to being the first person to import nutria to North America. Second, Ned McIlhenny was not the first nutria farmer in Louisiana, nor was he the first in the state to release nutria. Instead, Bernard's research suggests that Ned was at least the third Louisianan to maintain captive nutria, and that he was probably only the second nutria farmer in the state to purposely release his stock into the wild. Third, Ned did not import nutria from Argentina as he stated to Eugune Dumez; he purchased them from Bernstein's Louisiana-based nutria farm. Thus, as Bernard states in "M'sieu Ned's Rat?," the popular version of events regarding nutria introduction to Louisiana that blames Ned exclusively is somewhat bogus, omits pertinent facts, and is not wholly accurate.

M'sieu Ned's tall tales brought him undue blame for Louisiana's nutria invasion, but the fact remains that most folks would consider his approach toward nutria introductions to be cavalier. Although Ned seemed to provide A. P. Daspit with authoritative advice on nutria farming in 1930, he was brutally honest with Abe Bernstein. "I wish you would tell me how they should be cared for, as I know nothing about the care of these animals, and have never attempted to keep them in captivity," he wrote to Bernstein, before deploying Jungle Gardens landscape architect Jim Kennedy to receive the nutria.[54] Kennedy documented his trip meticulously, to the extent that he recorded an expense of $3.90 on food and gasoline, and provided Ned with a detailed, typewritten report of the transaction. Apparently, the trip did not go as planned. According to Kennedy:

Home away from home. Ned McIlhenny's nutria enjoy treats from a handler in their pen on the north end of Avery Island. Courtesy of E. A. McIlhenny Enterprises, Inc., Avery Island, Louisiana.

The animals were not in New Orleans, but at a point near St. Bernard. On arriving at the location it was found that the mother had 6 little ones and the fellow would not let us have them unless we paid the same for them as we did for the fourteen [adults]. An argument issued but was finally settled by Mr. Bernstein paying an additional 2 dollars each for the 6 babies.[55]

Kennedy also stated that Bernstein's instructions about maintaining the nutria were nonspecific and mysterious—feed them carrots, beets, cabbage, grass, alfalfa, or "anything green."[56] In other words, Bernstein was probably like Ned McIlhenny in that he knew very little about raising nutria.

With no curriculum for nutria care, a long process of trial and error followed the transaction with Bernstein. Bernard's research uncovered how M'sieu Ned reinforced his one-acre pen for the 20 nutria north of the Jungle Gardens after a few escaped. One nutria that was not recaptured was found killed 12 days later. "Nutrias [sic] have powerful teeth," wrote Ned to a potential nutria farmer that consulted him on raising the animals. He continued:

[The nutria] will cut out of any wood enclosing them, unless the wood is protected on the inside by wire-screen. They will cut the light poultry netting,

Nutria pens. Ned McIlhenny's nutria pens on Avery Island. Courtesy of E. A. McIlhenny Enterprises, Inc., Avery Island, Louisiana.

but will not cut heavy poultry netting. I use 2" mesh netting for protecting
the wood on the inside. Unless the pen in which they are enclosed has a floor
through which they cannot dig, the enclosure should be sunk at least 2½
ft.—preferably 3 ft.—into the ground. The pen in which I raise nutrias is large
... [consisting] of ordinary marsh, with holes dug in it, in which there is a
constant supply of water. The fence surrounding the nutria enclosure is sunk
three feet into the ground, and extends for four feet above the ground, and is
wood with wire covering. The wire covering, however, should not reach the top
of the fence, as nutrias [sic] climb where they have a mesh in which they can
get a foothold. They cannot climb up a plank; nor will they cut through a plank,
unless their hind feet are on the ground.[57]

Ned expanded the pen to about 23 acres by early 1939, with a maximum size of 35 acres by late 1941. Nevertheless, it remained challenging
to contain the population in a tight space. Another issue emerged when it
became apparent that the pen was not large enough to accommodate the
constant appearance of new kits. At first Ned enjoyed the fecundity, telling a friend, "They have eaten every spear of grass, and I am cutting the
heavy marsh grass with mowing machines and giving them a full truckload
every day, besides a little sorghum and corn-on-the-cob. They are thriving

wonderfully."[58] But it was only a few years before the 20 founding nutria became over 1,000, and Ned sold the kits to fur farmers all over the country rather than absorb the costs of feeding their voracious appetites.[59]

Ned was nowhere close to being the only person to sell nutria across state lines, although his transactions may have inadvertently compounded the issues of invasive nutria elsewhere. For example, biologists at Maryland's Blackwater National Wildlife Refuge, where officials are now close to eradicating nutria, bought a breeding male from Ned to pair with females already released at the refuge that hailed from as far away as Canada and California.[60] Nor was Ned alone in releasing unwanted nutria into the wild, which he did in June 1940 when he freed 21 nutria into the marsh surrounding Avery Island.[61] Some of these sold nutria were originally thought to have escaped through a compromised fence during an unnamed hurricane in 1940. But Bernard busted that myth with a memo Ned wrote to himself on June 1, 1940, a little more than two months before the storm, saying that he had "liberated" 21 nutria in the marsh on the south side of his "lower shooting pond," located about a mile northeast of the nutria pen.[62] Of course, Ned talked about his deliberate nutria releases in letters to the Department of Conservation with great pride. "You know my activity . . ." he boasted to state officials, "liberating them for the purpose of establishing an addition to our fur industry."[63]

In terms of this purpose, the nutria releases were a success. For years afterward, Ned sold breeding stock from his pen and local trappers took nutria from marshes on and around Avery Island. Ned also profited by sharing in proceeds from the fur of wild nutria trapped on his property. Perhaps the nutria endeavor was too successful, at least in the context of its era. The operation went so well that in 1945, Ned wrote to a nutria farmer in Washington State that he "let all of [his] nutrias go in the marshes." His original aim had been "to establish a fur industry on nutria in the waste marshes of Louisiana" and he believed that he had "succeeded in doing this."[64]

The amount of responsibility that Ned McIlhenny should bear for America's nutria problem remains debatable. On the one hand, Louisiana's burgeoning 1930s fur industry probably made it unimaginable that nutria and other furbearers would eventually replicate unchecked because of low hunting and trapping pressure, E. W. Nelson's warning letters notwithstanding. Nutria were also considered to be weed controls that were beneficial for the environment. Thus, to be known as the man who introduced nutria to Louisiana most likely would have been an enviable title. It also seems inappropriate to remember a man who supported environmental causes by building exotic gardens, saving snowy egrets, and setting aside coastal

Duck hunt. Governor Richard W. Leche hunting ducks during the 1930s at Avery
Island, Louisiana. Ned McIlhenny's balanced environmental philosophy meant that
he fought to save snowy egrets while operating a successful hunting outfit. Courtesy
of State Library of Louisiana.

marshlands that are now state wildlife areas for a well-intentioned but ultimately ill-fated side venture—even if Ned's own desire for celebrity placed
the blame for Dixie's nutria invasion squarely on his shoulders. On the other
hand, Ned McIlhenny started a nutria farm with minimal education on
nutria feeding habits or behavior, was aware of the risks, and sold nutria to
farmers in other parts of the country to create a revenue stream. However
well-meaning and unoriginal his nutria farm was, many folks feel that these
contributions to America's nutria problem are hard to deny—even if Brote,
Bernstein, Frakes, and others were also at fault.

Ned believed in a balance between conservation, sportsmanship, and
profitable resource utilization. Thus, one wonders whether Ned, were he
still with us, would shudder at the environmental harm nutria have caused
and repent his decision to farm nutria. The nutria pens were, after all, not
the first time that Ned's moderate environmental management philosophy
made him a flashpoint for controversy. In 1923, well before constructing

nutria pens on Avery Island, Ned apparently found no issues with opening a private hunting club for three months out of the year between two state wildlife refuges, especially because dues from the club supported the sanctuary. He once explained, "I have always been, and am yet, an enthusiastic duck shooter, but also . . . one who has an inborn love of birds, and a sympathy for them."[65] Ned was able to reconcile the two operations, but not everyone could do the same. Environmentalists dismissed his practices as duplicitous, calling the club a "death trap" for endangered wildfowl, and the hunting club enterprise ended abruptly.[66]

Ned's life, however, continued with a newfound vigor. During his final years, Ned guided McIlhenny Company through the Great Depression and World War II, and the 1945 liberation of "all of [his] nutrias . . . in the marshes" would be one of his last business decisions. Ned ceased his activity in the nutria trade that year and began lightening his workload after December 1946, when he suffered a stroke. After a few years of declining health, M'sieu Ned died in August 1949, and was buried on Avery Island under a succinctly written tombstone in the family cemetery: *Edward Avery McIlhenny. Born March 29, 1872. Died August 8, 1949.*[67]

Today, Ned's magnetism and love of the outdoors lives on. His Jungle Gardens and Bird City are major havens for bird and plant species, enjoyed by hundreds of thousands of tourists each year. Ned's illustrated and written documentation of Avery Island's fauna and flora was donated as a collection to Louisiana State University, and a collection of natural history books there is named in his honor.[68] And Ned's balanced philosophy remains part of Avery Island culture, with the McIlhenny family adopting the motto Man and Environment in Balance in 1971.[69] But some folks still remember Ned as a player in spreading nutria across the United States, and although previously accepted claims of the nutria problem being "all his fault" have been refuted by Bernard's research, Ned's fateful nutria farm remains a controversial part of his legacy.

Despite the nutria from more than half a century ago, the environment on Avery Island has thrived beautifully. After leaving the archives and driving through the sumptuous Jungle Gardens, I walk to the extreme north of the island, where Ned's nutria pens used to lie in a swampy area. I listen for the descendants of the Avery Island nutria, but hear nothing—they are gone now. And as the years come to pass, the arguments about how right or wrong Ned was about the nutria farm have also dissipated, like dew from grass on a hot morning in the bayou. What remain are the historical influences, both environmental and political, that caused nutria to not only proliferate, but also spread across North America.

# Chapter 4

# Alien Invasion

I can show you places where nutrias and muskrats grew so thick 40 to 50 years ago, trappers could not catch enough of them. They ate up all the grass, and those areas are now ponds.
—**Ignace Collins** in J. N. Felsher's "Last Days of the Trapper," *Louisiana Life* (2000)

It is fortunate that the nutria increased during a period of declining muskrat populations. Many trappers have thus been able to continue their trade of trapping in the marshlands. . . . Nutrias are also accused of competing with muskrats and preventing their increase.
—**V. T. Harris,** "The Nutria as a Wild Fur Mammal in Louisiana," Twenty-First North American Wildlife Conference (1956)

*Television can imitate society, and* Seinfeld *is no exception. The famous "rat hat" dialogue highlights the existence of niche markets for nutria fur, but the "Reverse Peephole" episode provides social commentary on the transition of wearing fur from a status symbol to social incorrectness. While Elaine, Jerry, and George attend the party of an acquaintance named Joe Mayo, Elaine becomes unhappy with her on-again, off-again car mechanic boyfriend David Puddy because he wears a fur coat. Mayo has a similar coat, and Elaine throws what she thinks is Puddy's coat out the window in a rage—the coat is actually Mayo's. But Elaine is reluctant to buy a replacement fur coat for Mayo because Jerry's nemesis (Newman the mailman) finds the discarded fur coat in a tree. Elaine asks for the coat back, but Newman has already given it to Svetlana, the wife of his landlord Silvio, with whom he is having an affair. When Jerry wears the fur as if it is his coat to keep their suspicious landlord from find-ing out about the affair—if Silvio finds out, he will surely evict Newman and*

*Kramer for installing "reverse peepholes" that look into their rooms instead of outward—the anti-fur sentiment among the otherwise insensitive characters becomes clear, with Silvio mocking Jerry as effeminate (". . . he's very fancy! Want me, love me! Shower me with kisses!" he says to imitate Jerry while curtseying) and Puddy calling Jerry "a bit of a dandy." Even Jerry himself agrees. "[The coat] is mine," he declares in an effort to have Elaine play along with the ruse. "I'm a fancy boy."[1]*

*The "dandy" and "fancy" comments are markedly different from the general sentiment of the 1970s when archetypal hero "Joe Willy" Namath, having led the New York Jets in a shocking Super Bowl upset after guaranteeing victory, took to wearing a fur coat on the sideline. And it is a far cry from the 1920s and '30s, when fur was a lifeblood industry in Louisiana and folks countrywide were starting nutria farms to replenish a dwindling fur stock that was exhausted by overwhelming demand. One can reasonably argue that Elaine's reaction to her boyfriend's fur coat demonstrates a desire for social conformity, and that furs are outside of the social conventions for the times. Some even suggest that Puddy's fur coat is a de facto symbol of feminism, because the actions of the characters suggest that a man wearing a fur coat is effeminate and socially incorrect. Either way, Joe Mayo's fur coat is not the only example of the negativity surrounding such apparel in* Seinfeld; *in episode 27, Elaine derides another woman for wearing a fur coat ("Is that a real fur? . . . You don't care that innocent, defenseless animals are being tortured so that you can look good?").*

*With these changing attitudes came falling prices for pelts and dwindling harvests of furbearing animals. Many would consider this progress, including the author. But this "progress" yielded an unexpected effect—it left nutria populations to expand and eat away at our wetlands unchecked.*

✦ ✦ ✦

The vast desert between Phoenix and Yuma, Arizona, is one of the hottest spots on the continent. Largely void of civilization and unmolested by alien nutria, the area seems like an odd side trip for our story. But surrounded by this desolate wasteland is a small town called Aguila where William F. Frakes, he of the California nutria farm, died in 1920 after his stock nutria were washed into a canyon over a decade earlier. Available records describe neither Frakes's activities in Aguila nor his reasons for assuming the community as his residence, but plenty of documents tell of another incident in this desert involving an introduced species.

Sixty-six miles west of Aguila rests the parched hamlet of Quartzsite, elevation 879 feet, which hosts a surprisingly well-known cemetery with an extraordinary attraction—a pyramid-shaped gravestone with a golden camel atop its pinnacle, standing stoically as a tribute to a fascinating but little-known part of American history. Around 150 years ago, camels (*Camelus* spp.) roamed these parts during a government-sponsored experiment called the US Camel Corps to use them as beasts of burden.[2] For a few years, the ability of the humpbacked ungulates to go without water made them useful pack animals between Tucson and Yuma; then came issues with their cranky dispositions and a more pressing national security matter called the Civil War. The Camel Corps lost its funding, and most of the camels were sold to landowners or released into the harshness of the Sonoran Desert.

The golden camel marks the grave of Ottoman immigrant Hadji Ali (later anglicized to Hi Jolly),[3] the operation's lead driver, who dabbled in freights, mining, packing, and scouting with his remaining camels. After failings in these businesses rendered him unable to maintain their care, Hi Jolly released his beloved camels into the desert before his death in 1902. Some believe that a final camel remains, unaware it is the last, roaming the desert to look for a mate that will never appear; others say that Hi Jolly's ghost wanders the desert in limbo, desperately searching for his herd. Either way, the last sighting of a camel in the area was in 1943, and an eerie plaque at Ali's grave offers insight to what could have been: "A fair trial might have resulted in complete success."[4]

Like the camels that were released in Arizona, the stories of nutria going wild in Louisiana are compelling and unusual. Both releases resulted from notions of using the animals for industry that seemed pragmatic and did not withstand the test of time. But unlike the camels, nutria became an *invasive* species, rather than dying out following their introduction. Most nonnative species never make the transition from *introduced*, or merely nonnative, to *invasive*—harmful to the environment. To achieve this dubious distinction, nutria had to be brought to the Gulf Coast, and also proliferate, spread, and cause ecological harm.

So why did *Myocastor coypus* take this extra step? The first answer is obvious. When rapid population growth and sweeping range expansion occurs, quick reproduction is often the catalyst. And then there is the number of introductions to consider—hundreds of released or escaped nutria formed populations all over the United States, whereas only Hi Jolly released camels. But for nutria and other invasive species, a unique story also explains the transition from nonnative to invasive.

European starlings, for instance, were introduced into Manhattan's Central Park for nostalgic reasons in 1890 by American Acclimatization Society (AAS) president Eugene Schieffelin. The society made dubious attempts to reintroduce every bird species mentioned in the works of Shakespeare into North America. Because of the AAS efforts, somewhat reminiscent of *La Société Zoologique d'Acclimatation*'s acquiring of nutria, European starlings now occupy nearly all of North America and the 100 or so initially released individuals have at least 150 million descendants that are devouring fruit, destroying crops, competing with native birds, and noisily squawking everywhere.[5] Another example comes from the six million feral pigs (*Sus scrofa*, also called *razorbacks*) that cause billions of dollars in property damage every year and are a potential vector for over 20 diseases of imminent interest to biologists and the public like brucellosis (*Brucella suis*), pseudorabies (*Suid herpesvirus*), and swine fever (*Pestivirus* spp.). Razorbacks are the result of folks releasing pigs into the wild to create game for hunters.[6] Kudzu (*Pueraria* spp.), an East Asian vine, was brought to North America as a form of erosion control and soil enhancement. But it became a noxious weed that has engulfed native plant life all over the southeastern United States.[7] And grass carp (*Ctenopharyngodon idella*), which have altered many aquatic systems by competing with native fish species for food and habitat, were imported to North America for aquaculture and emancipated through a series of accidental, illegal, or sometimes even authorized introductions.[8]

These introductions were well intentioned. But the consequences were misunderstood, and the releases poorly conceived. Shortsightedness also played a role in these stories. The organisms eventually outlived the introduction's original purpose, and the long-term issues with importation for aesthetics or environmental regulation were ignored in favor of short-term benefits. Such was also the case with nutria. Folks in the Gulf Coast region simply disregarded the possibility of a downward spiral in fur harvests that were already showing signs of unsustainability elsewhere on the continent. Even before the disinterest in fur reached its height after grassroots campaigns in the 1980s, some fur farmers released nutria when they proved to be a challenge to raise and failed to yield an immediate profit.

So why don't we talk about McIlhenny's nutria the same way we speak of Hi Jolly's camels? Partly because nutria are more fecund than camels, of course. But nutria-related damage would have been less salient if nutria dispersal had not been incited by state wildlife officials. As it was, folks who thought they were acting prudently by enacting a market shift from muskrat to nutria pelts wound up creating a bigger nutria problem to endure when fur markets declined. It is a cautionary tale—and hopefully, a lesson learned.

## Return of the Nonnative

While Ned McIlhenny was raising his nutria on Avery Island to address a specialized market, muskrats were abundant throughout Louisiana. Their natural predators were few, and the market for "'rat" pelts was going strong. The only blip on the muskrat radar was a flood in 1927 that lasted 108 days, resulting in intentional dynamiting of a Saint Bernard levee and the drowning of one to two million muskrats. The Department of Conservation was concerned enough about rebuilding populations in the muskrat-rich area around Delacroix Island so as to prohibit the sale and purchase of muskrat furs, and thousands of trappers took financial losses or relocated. But the long-term effects were minimal. Over five million muskrats were taken to the tune of approximately $8.5 million the following season. And that was without the production from Saint Bernard, which was down from its former yield but still produced more than 900,000 muskrats only two years removed from the flood.[9]

With the 1930s came tougher economic times in which fur products were not a priority. Fur sales plummeted during the Depression, as many folks no longer earned disposable income that could be used to purchase material goods. But folks still revered furs as a symbol of conspicuous consumption, and wealthy individuals who were minimally affected by the economic downturn bought luxury furs from high-end retail boutiques. In this altered market, Louisiana remained the center of America's fur industry and fared adequately during the Depression. The economic survival occurred not only because of the sheer number of available muskrat pelts, but also as a result of some euphemisms that marketed muskrat as an affordable alternative to other furs. Muskrat was sold as "Hudson seal," an inaccurate alias that reflected the popularity of seal furs, and eventually became "seal-dyed muskrat" after the influence of truth-in-advertising campaigns. When fashion interests shifted to *en vogue* furbearers like mink and away from seals, muskrats were coined "Southern mink" and "dyed muskrat." For $150, women could purchase coats with the look and feel of a mink coat worth $1,200.[10]

Fur markets continued to drop off during the World War II age of the early 1940s. With men going overseas, many families lost immediate income and were left with less money to spend on trivialities like furs. But Louisiana's economy dodged a bullet once again when the Department of Conservation intertwined muskrat trapping with national wartime efforts.[11] The federal government took control of the nation's natural resources, including the fur industry, and their campaigns through the Department

of Conservation encouraged those not fighting overseas to exploit Louisiana's fur resources and replace the $250 million in fur imports from China,
Russia, and Australia that were cut off by the war.[12] The Department of
Conservation even announced and supported business opportunities that
employed taken muskrats, including a bizarre program to use "marsh hare"
(another euphemism for *muskrat*) as a novel source of meat that never
progressed beyond preliminary discussions, a radical initiative to render
muskrat fat for the glycerin that was a key ingredient in explosives, and a
somewhat successful opportunity for trappers to sell "musk" scent glands to
perfume manufacturers.[13]

The good times for Louisiana muskrat harvesters continued deep into the
1940s, as the peak muskrat trapping years of the 1930s and '40s yielded over
four million muskrat pelts annually statewide. But muskrats still remained
numerous enough to generate reports of their "eating out"—destroying—
large areas of coastal three-cornered grass.[14] Perhaps the most famous
account came from Ted O'Neil, fur biologist for the Louisiana Department
of Wildlife and Fisheries (LDWF), who described the conditions in his seminal work from 1949 titled *The Muskrat in the Louisiana Coastal Marshes*:

> [An eat-out] occurs in the marshes when the muskrats have populated an
> area to the extent of completely eating the existing vegetation, including root
> systems which bind the soils together.... [In a complete eat-out,] the peaty
> floor is usually broken to a depth of as much as 20 inches. Soils of a marsh
> damaged in this manner disintegrate into loose muck, and floating in this
> ooze are decaying plant remains and occasional tufts of marsh sod that have
> been undermined by the feeding muskrats. The crevey [*sic*] that results usually
> remains for several years.[15]

Emboldened by the resilient 'rat populations, trappers took to the swamps
with alacrity. And so came the historical peak of muskrat harvesting in Louisiana, or anywhere else. During the 1945–47 seasons, often referred to as
the "Great Muskrat Eat-Out," more than eight million muskrats were taken
each year. Ironically, the most productive areas were the brackish marshes
of Saint Bernard Parish that were once underwater from the flood. Financial returns were awesome—for the 1945–46 season, the total fur harvest
was worth upward of $15 million, four-fifths of which was from muskrats.
This was followed by yields of over $9 million and $11 million for the 1946–
47 and 1947–48 seasons, respectively.[16]

In the midst of tremendous profits, no one seemed to realize a harsh
reality—the muskrat population was finally shrinking. At first, it was easy

Pond. A typical pond created in brackish marsh by a severe muskrat eat-out. The vegetation is winegrass (*Spartina patens*), and the area was eaten out in 1939, photographed in 1942, and revegetated in 1947. Marshes broken to this extent generally remain unproductive for long periods unless artificially drained. Courtesy of Louisiana State University Special Collections; T. O'Neil, *The Muskrat in the Louisiana Coastal Marshes: A Study of the Ecological, Geological, Biological, Tidal, and Climatic Factors Governing the Production and Management of the Muskrat Industry in Louisiana* (New Orleans: Louisiana Department of Wildlife and Fisheries, Fur and Game Division, 1949).

for trappers and state biologists to dismiss the sudden drop in muskrats as a routine down cycle, as muskrats usually exhibit up-down population cycles of around 10–14 years.[17] But after two to three years, it became clear that the situation was unfolding differently than the usual cyclic changes or vacillations. As state officials decided their course of action, trappers were faced with the choice of focusing on other furbearers as a segue to the next upturn in the muskrat cycle, or leaving trapping altogether for other employment in oil drilling or gas extraction.

By 1949, the entire Louisiana fur industry, which was largely dependent on an influx of muskrat pelts, was profoundly and unexpectedly at risk. With the state's economy hanging in the balance, LDWF officials scrambled to publish a series of articles in their magazine, *Louisiana Conservationist*, attempting to reassure anxious trappers that the muskrat downturn was fleeting. The main hypothesis proposed for the "temporary" slump in muskrat populations with these publications was a severe drought that coin-cidentally occurred at the same time as the natural downtick that muskrat populations experience every 10 years or so. Other potential reasons included an increase in raccoon populations. The price on raccoon pelts fell precipi-tously to one-tenth of a muskrat's fur during the 1940s, and between this and the larger size of raccoons, which necessitated more lifting and skinning effort and could only be done on a few animals at a time, raccoon popula-tions were left to multiply without control. Muskrat kits were predated by raccoons, said the articles, causing millions in lost revenue for trappers.[18]

Maybe the housing development or the oil, sulphur, and gas industries were to blame for the low numbers, claimed the LDWF. Drills and marsh buggies used during geological exploration may have created outlets for saltwater to creep into muskrat habitat. Or perhaps muskrats were declin-ing because of the construction of levees along the Mississippi and Atcha-falaya Rivers for flood protection. The articles mentioned how this made people feel better about living on the floodplain, since the river was now unlikely to overflow its banks. These disruptions likely increased the salinity of delta waters, and the three-cornered grass enjoyed by muskrats was being replaced by more salt-tolerant grasses.[19] And muskrats, according to the LDWF, did not help their own cause. They created open water by widening the holes started by digging exploration crews, which increased tidal ero-sion and decreased the amount of their precious freshwater marsh habitat.[20]

Maybe it was a combination of these reasons, but in any case, Louisiana's fur industry was on the line; the number of trappers fell by around 40 per-cent just before the 1950–51 season. Meanwhile, postwar Europe flooded the market by exporting fur across the Atlantic in droves, leaving the market in transition and fur biologists shaking their heads.[21] State officials needed a quick boost for their in flux fur industry. Thus, a stage was set for the nutria already imported to satisfy niche markets. Instead of being relegated to captive fur farms, nutria would replace muskrat as the primary game for Louisiana's trappers.

Ned McIlhenny began releasing nutria into the wild in 1940. But sev-eral other nutria farms freed nutria or let them escape from enclosures during the late 1930s and early 1940s. These nutria releases were usually seen as a welcome addition to muskrats, and the populations proliferated quickly. This probably explains why nutria were found west of Avery Island in Lake Arthur during the same year of Ned McIlhenny's first release. Per-haps Ned's first round of releases account for the nutria colony in Vermilion Parish found in 1942, just 30 miles away from his Jungle Gardens, in the freshwater marshes north of White Lake. Other newfound populations of

Trapped. A nutria in a trap. Courtesy of Louisiana State University Special Collections: Donald W. Davis Slide Collection, Louisiana Sea Grant Collection Images, Louisiana Digital Libraries.

nutria in western Louisiana were probably started unintentionally. Nevertheless, the early 1940s were peppered with reports of vigilantes trapping and transplanting feral nutria into marshes from Port Arthur, Texas, to the Mississippi River.[22] More famously, the winds of a 1941 hurricane in Texas are usually blamed for dispersing nutria in southeast Texas and southwest Louisiana.[23]

In any case, with scattered populations of nutria throughout the Gulf Coast region, nutria began to occupy traps that were meant for muskrats. Eventually, trappers came to see this as an opportunity rather than an inconvenience. By 1941–42, nutria were the focus of trapping operations on the Sabine and Lacassine National Wildlife Refuges in western Louisiana instead of muskrats. During the next year, quickly growing nutria populations became well established throughout most of the western Louisiana and east Texas coast.[24]

State wildlife officials also viewed nutria as an opportunity. Even with the highest ever muskrat numbers in the early 1940s, the LDWF figured that nutria would coexist with muskrats despite their similar ecological niches. And because it was then inconceivable that furs would become unmarketable, the LDWF assumed that trapping would keep the combined numbers of nutria and muskrats at bay. Having already communicated with Ned McIlhenny about releasing nutria as an addition to a flourishing fur industry, they commenced the first nutria trapping season in 1943–44, resulting in a paltry take of 436 pelts.[25]

Meanwhile, land operators that were replacing freelance trappers wanted a piece of the pie and purchased live nutria from Vermilion Parish for transplanting into their marshes. Thus, nutria populations increased and so did take, which was listed as 8,784 nutria at a price of five dollars each for the third official nutria season of 1945–46—all by a single company, which acquired nutria well into the 1950s.[26] By this time, nutria were already found at the Delta National Wildlife Refuge (DNWR); on the extreme eastern edge of Louisiana at the mouth of the Mississippi River; in Saint Mary, Iberia, Vermilion, and Cameron Parishes; and most of the coastal area in between.[27]

Prior to the 1945–46 season, the number of nutria trapped numbered only several hundred. But with nutria catching on as game, the state legislature enacted Act 197, which gave the LDWF authority to officially protect nutria by giving them the same December–February trapping season as muskrats and placing a 10-cent severance tax on each nutria pelt taken.[28] Now it was on paper—nutria were an official furbearer in Louisiana, and their take doubled to the tune of over $54,000 in the year following the legislation.[29] For the next five seasons, the price per nutria pelt remained over three dollars and take rose to more 78,000, valued at approximately $364,000.[30]

When muskrat numbers went down later that decade, the already-dispersing nutria provided an increasingly available alternative source of fur. The LDWF hailed nutria presence as a "Godsend" for the state's economy, strangely describing the nonnative rodents that were being killed en masse as "docile and likeable rodent[s]" that could become not only the state's richest fur crop, but also maybe even "the most valuable fur producer on the North American continent."[31] The message was clear. Until the late 1940s, nutria fur played second fiddle to muskrat. Now, nutria was taking the lead.

As nutria multiplied and became ensconced in Louisiana's marshes, the LDWF facilitated a market transition from muskrat to nutria by transplanting and protecting nutria. These stocking efforts during the 1940s catalyzed an explosion of nutria populations. But some of the most profound nutria dispersal came when state officials promoted nutria as a biological agent for the control of aquatic weeds, primarily water hyacinth (*Eichhornia crassipes*) and alligator weed (*Alternanthera philoxeroides*). Nutria were

Trappers. Two nutria trappers sit outside while the skins dry on the rack. Courtesy of Louisiana State University Special Collections: Louisiana Department of Wildlife and Fisheries Slide Collection.

Drying out. A rack of drying nutria pelts, in Bayou DuLarge, Terrebonne Parish, Louisiana. Courtesy of Louisiana State University Special Collections: George Castille Slides, Louisiana Digital Library, Baton Rouge, Louisiana.

also advertised as providing these ecological benefits while being incapable of causing eat-outs, which was a misconception because they could at high population densities. Get-rich-quick confidence men took note and sold nutria as "aquatic weed cutters" to a public that knew no better. Upon receipt of the nutria, the rodents cut weeds and most other vegetation as well, reproducing faster than their keepers could feed them. It complicated matters that nutria tended to roam throughout their range, unlike muskrats, which were generally philopatric and stayed closed to their burrow. Between the insincere promoters and the misguided state officials, nutria were translocated into Alabama, Arkansas, Georgia, Kentucky, Maryland, Mississippi, and Oklahoma. Nutria also moved deeper into Louisiana and Texas.[32]

Further vagaries in the fur market probably contributed to the nutria diaspora. For example, the mink coat was no longer a fad when the 1950s arrived. It was all the rage, and when folks sought the lustrous cloak of *Neovison vison* rather than nutria, some trappers refocused accordingly. Another issue was that the market for southern Louisiana pelts shifted to garment makers in Europe. Five years later, when the growing number of pelts from increasingly copious nutria exceeded their demand, nutria prices plummeted to just one dollar per pelt and incentive to harvest nutria was reduced.

Even with these variations, the overall trend was that translocations of nutria continued and take increased—yet somehow, nutria populations still burgeoned. Meanwhile, translocations became increasingly cavalier. One of the most egregious and poorly conceived transplants of nutria occurred in 1951 when the Fur and Refuge Division of the Wildlife and Fisheries Commission plopped 250 nutria into the Pass-a-Loutre Public Shooting Grounds, located in the heart of southeast Louisiana's Mississippi delta. The intent was that nutria would control undesirable vegetation and enhance trapping opportunities.[33] Before this event, practically all nutria trapped in Louisiana were caught in the southwest or south-central portions of the state. Afterward, the geographic range of nutria was artificially expanded into one of the few key ecological areas that was not previously occupied by nutria. The incident seemed to publicly communicate the imminence of a market switchover from muskrat to nutria. It was not long before folks began to implicate flourishing nutria populations in the decline of muskrats, a falloff that was completed during the 1961–62 trapping season, which was the first season when more nutria (over one million) were taken than muskrats (around 600,000). Indeed, the harvest of muskrats never again exceeded that of nutria.[34]

Pass-a-Loutre. The 66,000-acre Pass-a-Loutre Shooting Grounds near the mouth of the Mississippi River, shown here during the 1940s, was operated by the LDWF and was the site of an infamous nutria transplant. Courtesy of State Library of Louisiana.

Harvest of muskrats tended to decline as nutria populations increased. Public perception was that nutria populations were causing the muskrat declines, but whether that was reality remains unclear. Indeed, historians continue to disagree on whether nutria negatively affected the muskrat populations, and if they did, to what degree. Proponents of the notion that nutria minimally affected muskrats argue that the two species required different habitats and direct food competition was reduced. Whereas muskrats preferred brackish marshes that were resistant to flooding so that they could build their mounds, nutria favored freshwater and were less picky about the location of mound construction. In addition, nutria were probably more adaptable than muskrats because they were more likely to search for a new home when necessitated by less than optimal environmental conditions.

But critics of that hypothesis typically favor the reasons proposed by the LDWF for the muskrat decline, maintaining that nutria were not as susceptible to drought and were better able to defend against raccoon attacks.[35] Indeed, muskrat populations have been found to skyrocket in good conditions while being inordinately susceptible to climatic catastrophes involving precipitation—rain can drown muskrats in their burrows.[36] Thus, as 'rat populations dwindled, the state was adamant in their conclusion that "the nutria is no menace to the muskrat," and that there was plenty of marshland to sustain both rodents.[37]

It is also possible that muskrats began a natural decline that was exacerbated by the presence of competing nutria, leading to what has been called the "fall of the muskrat."[38] Perhaps 30–40 million muskrats, 14 million of which were taken, were chomping away at Louisiana marshes during the Great Muskrat Eat-Out of 1945–47. Because habitat did not rebound after the eat-out, it seems unlikely that lack of habitat alone can explain the steeply collapsing muskrat population. Industrialization may have also interacted with a series of coincidental circumstances, such as normal marsh erosion and subsidence, severe storms and prolonged summer dry cycles, and the proliferating nutria.[39]

The debate over how much nutria caused the muskrat decline continues to this day. But either way, the result was the same. Muskrats went into decline, nutria replaced muskrats and mink as Louisiana's *en vogue* furbearer, and nutria populations peaked at an estimated 20 million during 1955–59.[40] Once concerned about not having enough game for trappers, state biologists suddenly were left with too many nutria to manage. The same applied to Louisiana's trappers—some were already moved on to other industries after muskrats went into decline, and remaining trappers were helpless against millions of nutria. Thus, as the nutria populations swelled, environmental damage became imminent.

## The Outlaw Quadruped

While interested parties were considering how to handle mushrooming nutria populations, the rodents were eating away at the swamps. All of Louisiana felt their wrath, especially the delta region, where state biologists transplanted nutria during the 1940s. Many formerly dense stands of cattail (*Typha* spp.) were transformed into eat-outs—areas of open water. To make matters worse, nonnative plants and native plants that were inedible to local fauna sometimes filled an eat-out, rendering the structural and ecological integrity of the marsh unstable. According to Ted O'Neil:

> Two of the most intensively trapped and managed properties, on a tract of
> 150,000 acres in Vermilion Parish and a tract of 155,000 acres in Cameron
> Parish, were completely leveled by nutria, with peak takes of marketable pelts
> in the 60,000 to 70,000 figure, after which came sharp drops in production
> accompanied by poor pelt quality due to the lack of food supply. The active
> Mississippi River Delta, comprising about 350,000 acres . . . also went to pieces
> by 1956–57, as a result of the 250 nutria transplanted there in 1951. There was a

Seeking shelter. Nutria seeking higher ground after Hurricane Audrey on Pecan Island in 1957. The image is looking west down the road on Pecan Island about 4,000 feet from Highway 82, and the water in the marsh is higher than normal. Courtesy of State Library of Louisiana.

gradual rise [of the population] from 1951 to the peak in the 1955–56 season, when nutria were everywhere with vegetative cover still standing. By the following season, only the pass banks appeared to be holding the delta together.[41]

Things would get even worse for the Gulf Coast region following the 1956–57 season. A few months later, in June 1957, the eye of Hurricane Audrey made landfall in southwestern Louisiana where nutria were dispersed by the unnamed hurricane 16 years before. Marsh structures that may have ordinarily prevented the storm surge from inundating interior marshes did not because they were destroyed by nutria. With a reduced buffer between the massive ocean waves and the coast, damage from the surge extended 25 miles (40 kilometers) inland.[42]

Before there was Katrina, there was Audrey. As a category 4 hurricane with sustained winds of 145 miles per hour (230 kilometers per hour), Audrey would have been quite destructive even without nutria impairing the coast's reliability. Causing $147 million worth of widespread property and infrastructural damage, Audrey was at the time the fifth-costliest hurricane in the United States since 1900. Offshore oil facilities suffered $16 million of this damage, with a drilling rig sinking and four others pulled away from the ground. Audrey also killed at least 416 people, mostly in western Louisiana and eastern Texas, and continued inland into the upper South and Midwest after striking the Gulf Coast, with reports of tornadoes, death, and destruction as far north as Quebec.[43]

It was hard to find a bright spot in the aftermath of Hurricane Audrey. But given the escalating situation with expanding nutria populations, most folks found it useful that many nutria were drowned in Audrey's storm surge. Even though nutria disperse to new areas when flooding occurs, the area over which Audrey's damage occurred was too large for nutria to remove themselves, and their populations proved to be susceptible to the prolonged flooding. An estimated 60–65 percent of the nutria in the White Lake and Grand Lake marshes of southwestern Louisiana perished during Audrey's wrath, and probably over two-thirds of the nutria at Marsh Island were dispatched as the storm moved through.[44]

State biologists were hoping that the nutria problem was solved by Audrey. But mortality among nutria from the hurricane did not provide the desired relief. In fact, Audrey exacerbated the nutria invasion considerably by dispersing significant numbers of nutria inland, directly into the state's agricultural crops, such as rice and sugarcane. Damage to the wetlands somewhat subsided because there were fewer nutria overall, and nutria near the coast where the hurricane hit at full strength were most likely to perish. But nutria were now widespread over most of Louisiana, and crops were a new forum for creating environmental and economic disruption. In terms of damage, it was six of one and a half dozen of another.

Almost immediately following Hurricane Audrey, reports of agricultural damage increased. Rice was the most drastically affected crop, and nutria were found to live in rice fields year-round if not controlled. Fields of rice that were flooded with water from routine storms attracted the semiaquatic nutria, which used them as swimming grounds and constant food supply. Nutria remained in fields as the rice crop developed and until the rice was harvested, usually feeding at night. The foraging behavior of nutria meant that more plants were destroyed than eaten because nutria utilized the rice stems for only about five inches above the ground, and sometimes killed rice stems when they cut them off below the water level. Even if the stems were cut off above the water, destruction still occurred because this slowed or prevented the growth of a mature seed that was ready to be harvested when the field was cut.[45]

The most pronounced damage to rice fields occurred not when nutria overtly ate the rice, but when they burrowed into large irrigation canal levees. Muskrat burrows were already commonplace near these levees, but nutria carved out a much larger tunnel or enlarged an already existing one from a muskrat. Larger burrows meant an increased chance of a break in the levee. Running water near pumping installations in rice fields also attracted nutria, and these issues were compounded when cattle grazing on large levees stepped into nutria burrows and either enlarged them or created another problem for the farmer by becoming injured.[46]

In contrast to rice fields, where nutria usually dropped anchor after they discovered the arrangement, nutria seemed to only visit sugarcane fields and then move. But nutria damage to sugarcane was still profound because nutria damaged mature canes by completely cutting their stalks or completely uprooted young canes with their profligate eating habits. If the sugarcane field was protected by a levee, nutria sometimes burrowed into the levee and cut it, allowing the field to be flooded and causing a reduction in cane yield. Other crops damaged by nutria after they were driven inland by Hurricane Audrey included corn, beets, grain sorghum, peanuts, melons, sweet potatoes, cabbage, lettuce, and peas.[47]

Once a "Godsend," *Myocastor coypus* became a fugitive. Owing to their effect on agricultural interests, the Louisiana legislature took nutria off the "protected species" list and declared *M. coypus* an "outlaw quadruped" in 1958. A 25-cent-per-tail bounty was authorized, although the funds were never appropriated.[48]

## The Fur Flies

As Louisiana's fur industry entered the 1960s, an interesting paradox developed regarding public attitudes about trapping. With an unpopular war in Southeast Asia and bra burnings to protest traditional female societal roles, the decade was a rebellious period during which people became disenchanted with the establishment and were comfortable with political action. Along these lines, the '60s brought increased environmental awareness and more concern for animal rights. Nationwide, a growing concern for wildlife discouraged wearing furs, causing pelt prices to decrease. But somehow, the decade also saw an upsurge in furbearer harvests, in spite of various protests against using animals for fur. Perhaps this was because harvest numbers were left with little place to go but up after a trapper exodus to the oil and gas industries during the 1950s.[49]

Widespread public approbation of the anti-fur movement was novel for the 1960s. But the anti-trapping component of this campaign was four decades old, with crusades beginning during the Roaring Twenties. Often conglomerated with the current anti-hunting movement, the anti-trapping movement began, ironically, when fox and raccoon hunting organizations became angered by accidental trapping of their hunting dogs and incensed by competition with trappers. The movement's thrust was that commercial trapping severely depleted North American mammal populations, but the initiatives also included humanitarian objectives. For example, the Anti-Steel Trap League (ASTL) was one of the first organizations to focus specifically on the issue of wild animal trapping and the question of what constituted "humane" take. In the 1920s and '30s, the ASTL successfully campaigned for bills banning or restricting the leghold trap and other "inhumane" trapping practices in at least five states.[50]

These laws were repealed during the Depression in the 1930s, because the public majority believed that furbearers were needed for food and much-needed income. But via the state legislative victories, the anti-fur and anti-trapping positions were established and the arguments against trapping stated. In 1947, anti-trapping activists gained even more momentum when the Defenders of Furbearers (DOF) was founded. Later renamed Defenders of Wildlife (DOW), the organization educated the public about trapping and dissuaded folks from wearing fur coats that were popular during the 1930s. DOF was partially responsible for the national plummet of fur sales to an all-time low in the 1950s.[51] The Humane Society of the United States (HSUS), formed in 1954, is another nonprofit organization that generated national awareness of the fur trade and animal welfare issues during this period.

But trapping was a salient part of Louisiana's economy, and the same compelling arguments about sustenance and supplemental income used during the Depression continued applying to the state's residents. The need to control feral nutria from accidental releases and planned transplantations therefore trumped the anti-trapping campaigns of activist groups, and the trapping industry continued growing. When nutria permanently overtook muskrats and mink in take and value during the 1961–62 season, the LDWF started capturing nutria from one area of the state and releasing them in another to spread the wealth of the harvest.[52] The relocations satisfied their intended purpose, making nutria populations more ubiquitous than in the 1950s and providing nearly limitless stock for potential trappers.

Damage to agriculture was the trade-off of increasing nutria popula-tions, and this necessitated dispatching nutria more than ever. When the LDWF searched for new markets interested in nutria products, they found one—postwar Germany. Deutschland was already a player in the interna-tional nutria trade going three decades back. It was, after all, Canadian C. R. Partik's unnamed German friend who sent him a few nutria in 1928, leading to the publication of a pamphlet on how to raise captive nutria that encour-aged the advancement of nutria farming. By the 1960s, the nutria market in Germany blossomed, and their fur market began importing over one mil-lion nutria pelts per year.[53]

The arrangement between the LDWF and Germany seemed to achieve its objectives. With the renewed economic benefits of trapping nutria, their annual harvest climbed steadily during the early 1960s. And relatively fewer nutria led to less frequent complaints of nutria damage to crops. But even with increased harvest, transplanted nutria were multiplying so quickly that they were still causing extensive agricultural damage to sugar crops in coastal parishes. Such was the foundation for conflicts between fur and agri-cultural interests, particularly the sugar industry. Sugarcane, like fur, enjoyed an illustrious history in Louisiana and was going on 200 years as an inte-gral part of Louisiana's economy. Originating in New Guinea, sugarcane was carried by Christopher Columbus from the Canary Islands to Hispaniola during his second voyage to the New World in 1493. The sweet-tasting crop became popular in Louisiana after Jesuit missionaries carried sugarcane plants from Saint-Domingue, a sugar-cultivating French colony on Hispan-iola, to a current-day downtown New Orleans church on Baronne Street in 1751. Jesuits produced the first record of successful cane production with this translocation—technically, sugar was probably first planted in Louisiana during the late 1600s by d'Iberville, he of the aforementioned expedition to locate the mouth the Mississippi River. D'Iberville planted sugarcane from Saint-Domingue along the lower Mississippi, but this attempt to cultivate it failed.[54] In any case, warm weather in New Orleans provided the necessary conditions for sugarcane to grow, and folks loved the plant's sweetness and chewability. By the late 1750s, a sugar mill was built by a man named Claude-Joseph Dubreuil de Villars who lived on Esplanade Street.[55]

A key event in the history of Louisiana's sugar industry occurred when sugar plantations in Saint-Domingue were destroyed following a 1791 vio-lent revolt of slaves against French planters. Some colonists fled Hispaniola to seek asylum in New Orleans, and those who were experienced sugar-makers brought their skills along. Étienne de Boré, a prominent French planter who was later appointed the first mayor of New Orleans during the

US administration of the city, became the first person to produce economically successful raw granulated sugar in the area during 1795 after hiring
one of these refugees, named Antoine Morin, at his wife's family's plantation near present-day Audubon Park.[56] Marrying the daughter of Jean-Baptiste Destrehan, the former state treasurer, and risking their fortune in the
enterprise, de Boré made 100,000 pounds of sugar and molasses valued at
$12,000, changing Louisiana sugarcane from a few canes in a church to a
profitable commodity crop.[57]

Other entrepreneurial colonists followed in de Boré's footsteps and produced sugar commercially, usually buying the cane they planted from Saint-
Domingue or the Jesuits. Folks from other areas of the South heard about
opportunities associated with the sugar industry and came in droves after
the United States finalized the Louisiana Purchase in 1803. The quantity
of sugar production increased even further after wet weather and insects
decreased the indigo crop, which was the most important plant for many
area farms. Sugar eventually overtook indigo as Louisiana's favored cash
crop, increasing to a harvest of around 300,000 tons per year, with 75 operating mills employing half a million Louisianans.[58]

But sugar markets endured the same type of fluctuations and industry
dynamics as the fur industry. In fact, the sugar industry became a significant force in Louisiana's economy only after some challenges in developing
the industry were met. The War of 1812 slowed development of the sugar
industry, as did concerns that the subtropical climate and sea-level elevation of New Orleans was less than ideal for cultivation of a tropical plant
that grew on mountainsides. It was not until two new imported frost-resis-
tant strains alleviated this concern that sugarcane harvesting expanded outside of New Orleans proper. Then, a still-in-use device invented in 1834 by
a Paris-educated, New Orleans free man of color named Norbert Rillieux
called the *triple-effect evaporator* made sugar production more efficient,
as did the practice of slavery, which confined hundreds of thousands of
"colored" individuals by the 1860s.[59] The latter was overturned by the outcome of the Civil War, which removed most slave labor and drove sugar
output into a 50-fold reduction during the transition from a slavery-based
economy to a wage-oriented system.[60] As the sugar industry regrouped during the next few decades, mechanization techniques driven by animals and
then by steam, electricity, and gasoline were adopted.

Even with the benefit of technological advancements, labor was a constant problem for the sugar industry after the Civil War. Despite the advent
of sugar rations, the issue was exacerbated by men leaving to fight in World
Wars I and II. The Great Depression did not help the sugar industry either,

as folks spent more conservatively and it became less valuable to own a farm. These events of the mid-twentieth century created another period of turmoil from which the industry was recently emerging. The last thing sugarcane growers associations needed was a bunch of hungry rodents with long orange teeth to ruin their product as the industry reasserted itself. To sugar barons, nutria were pests, not "protected wildlife." But the nuisance status of nutria with regard to the sugar industry was anathema to LDWF efforts that promoted nutria as a wildlife resource.

The LDWF finally broached the conflict in 1963, when they declared nutria an "outlaw quadruped" in 17 coastal parishes. From 1963 to 1967, a nutria damage control research program evaluated trapping and shooting as nutria management techniques. The program helped negotiate a 1965 compromise between agricultural and fur interests. Under the arrangement, nutria were given statewide recognition as a protected furbearer, and returned to the protected wildlife list unless they were on agricultural lands. Thus, control of nutria deemed as agricultural nuisances by farmers was legal, and no permit was necessary.[61]

In the years following this arrangement, nutria harvest levels increased and the 1960s ended with a decadal nutria harvest of almost 12 million.[62] The healthy nutria take meant reduced complaints of damage to sugarcane or other crops, with issues usually minor enough to be remedied with localized trapping or shooting of nutria in the affected fields. Rice production eventually shifted to underground irrigation, which removed the opportunity for nutria to affect yield by chewing into pipes. Negative effects of nutria feeding in fields also decreased after millions of nutria were killed during a severe freeze event involving temperatures as low as 12°F (−10.4°C) in February 1962, just before the nutria damage control research program began. Trappers who saw nutria up close reported that surviving nutria sometimes lost their tails or feet to frostbite. And Hurricane Betsy, a category 4 storm that struck the coast in 1965, reduced nutria populations significantly—an estimated 400,000 were dispatched by the storm, along with around 70,000 muskrats.[63] But with so many nutria across the Gulf Coast region, these losses were a drop in the bucket. Like the sugar industry, nutria continued to bounce back.

## Reaping the Harvest

The anti-fur movement was coming into prominence as the 1970s rolled along, but its philosophy was not yet mainstream. A large group of

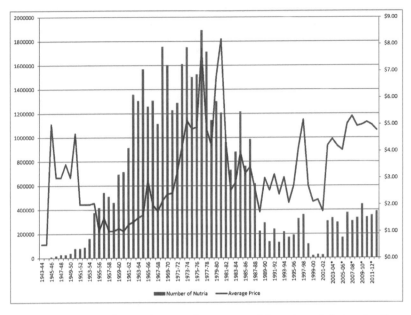

Price chart. Number of nutria harvested per year and the prices of their pelts. Courtesy of the Louisiana Department of Wildlife and Fisheries.

influential people remained interested in wearing fur. Nutria pelts were sometimes a sought-after luxury fur worn by style icons like Greta Garbo, Elizabeth Taylor, Sophia Loren, and Ursula Andress. The popularity of all furs also remained significant as celebrities like Diana Ross, Farrah Fawcett, and Raquel Welsh continued to wear furs on screen and in public events. Men enjoyed wearing fur too, as demonstrated by Joe Namath's propensity for fur coats on the sideline. Thus, the market for furbearer pelts was excellent, with the value of all pelts taken in Louisiana near $25 million for just one season (1975–76).

Furs like mink and fox continued as marquee products, but LDWF's wildlife management scheme during the previous two decades meant that most of the harvest value came from nutria. Spread over almost the entire coastal part of Louisiana, nutria multiplied to create supply, and trappers took advantage of the international demand. A network of 130 local buyers and 35 dealers formed, with about 99 percent of the nutria pelts acquired from trappers being shipped out of state.[64] Buyers and dealers acquired nutria pelts from Louisiana harvests that averaged about 1.5 million per year during the 1970s, with a peak harvest of 1.8 million pelts worth $15.7

million during the 1976–77 season. Indeed, total harvest of nutria for the 1970s topped an almost inconceivable 15 million individuals.

Europe usually received nutria pelts from dealers in Louisiana to make coats, but other uses were becoming popular. Nutria were sometimes used for mink food on farms that were producing the weasel-like animal for stylish coats, and nutria teeth were used by some Louisianans as "swamp ivory" to make jewelry.[65] Another creative use for ever-growing nutria populations came from the US Department of Agriculture (USDA), which used nutria meat to grow screwworms for the development of laboratory experiments that identified parasites in farm animals.[66] Taken together, these uses resulted in fewer reports of nutria damage to agricultural lands, an intended effect of initiatives taken by the Department of Conservation during the 1960s. It seemed that Louisiana's fur industry, with the help of fruitful nutria and some clever negotiation with the state's sugar moguls, had seen the storms through—sometimes literally.

To a certain extent, the positive national attitude toward fur continued into the early 1980s, as fur fashions continued to be a symbol of opulence displayed by actresses Joan Collins and Linda Evans on the popular television show *Dynasty*. Again, nutria was not the most sought-after fur, but was nevertheless in demand to the tune of $8.19 per pelt during the 1980–81 season. But the value of nutria and other mammalian pelts would not last much longer. Behind the steadily increasing sales levels during the 1970s and '80s, well-organized anti-fur campaigns were finally gaining enough steam to have a profound negative effect on the fur industry. Fur wearing was becoming politically incorrect. Indeed, anti-fur campaigns and advertisements became embedded in American popular culture via one of the same routes that made pro-fur sentiments popular—celebrity backing. They also flourished because of creative, high-visibility campaigns that promoted anti-fur viewpoints to a mass audience.

The anti-fur movement gained even more momentum during the anti-sealing campaigns of the 1970s, when protesters vehemently abhorred the clubbing of baby seals. Campaigns eventually widened their focus to all animals that were under siege because of the garment business and strongly discouraged consumers from purchasing fur products.[67] Perhaps the most famous example of a celebrity involved with this movement was French actress and model Brigitte Bardot, who protested seal hunts in Canada and promoted anti-fur sentiment during the 1970s. Many folks remember the poignant print materials from the campaign, which featured photos of cute, furry seals cuddling Bardot's chin juxtaposed over bludgeoned seals scattered along the coast of Labrador.[68]

Consumer opinion on furs continued evolving with the emergence of Peter Singer's seminal book *Animal Liberation* (1975), which introduced the concept of *speciesism*—discrimination against particular species—and thrust "animal rights" issues into the national spotlight.[69] *Animal Liberation* was an inspiration to two young activists named Alex Pacheco and Ingrid Newkirk, who in 1980 founded People for the Ethical Treatment of Animals (PETA), probably the most well-known animal rights group in the United States. PETA's mission became even more topical a few years later during the famous Silver Spring Monkey Trial of 1981, which shed light on the issue of animal experimentation. Named after the town in Maryland where the National Institutes of Health (NIH)–funded experiments took place, the case revolved around the laboratory of Dr. Edward Taub, who was conducting electroshock therapy on 17 monkeys kept in small wire cages and unsanitary conditions. Experiments took place under the pretext of studying nerve regeneration, but after the undercover work of Pacheco, Taub became the first person working in a lab situation to be convicted of animal cruelty charges.[70] The monkeys were eventually euthanized or died while in NIH care after PETA campaigned for their legal guardianship. Nevertheless, the incident provided PETA with national visibility and probably signaled the birth of the modern animal rights movement, which included a substantial anti-fur component.[71]

As PETA gained notoriety, other animal rights groups made an impact. For example, Lynx was a British animal rights organization that helped make the anti-fur movement part of popular culture. Targeting middle- or upper-class white women, Lynx produced gaudy print advertisements to destroy the public's perception that wearing fur was a status symbol. Arguing that wearing fur was wrong on a purely moral basis, Lynx advertisements showed animals caught in steel traps. To overtly shame those who wore fur products, Lynx produced memorable and influential slogans, such as "It takes up to 40 dumb animals to make a fur coat . . . but only one to wear it" and "Rich bitch . . . poor bitch . . . ," the latter complete with a picture of a woman wearing a fur coat next to a dead fox.[72]

In the United States, PETA was taking a similar approach as Lynx and using visually compelling advertisements to demonstrate their point. Their hallmark was the notorious, attention-getting, "naked" print campaign, marked by the slogan "I'd rather go naked than wear fur." Posters exhibited models and other celebrities photographed with only their arms and legs to cover themselves. Most of the posters featured unclad women, but occasionally well-built men like basketball player Dennis Rodman or rock musician Tommy Lee posed au naturel. Meanwhile, freelance activists gained national attention by pouring red paint on women wearing furs in San Francisco's Union Square.[73]

Although they were largely effective, the campaigns of organizations like PETA, Lynx, and others were not without criticism. For example, when pro-environmental, nongovernmental organization (NGO) Greenpeace International (GI) announced their aggressive media campaign against wearing fur in 1984, the resistance was substantial. Folks were concerned about the effects of the campaign on indigenous trappers in northern Canada, and the plans were canceled.[74] The unorthodox and barefaced methods of PETA and Lynx were also heavily criticized for crude language and grisly images that some thought were unnecessary. A few people even considered the posters to be sexist, or were offended by what they thought was a vulgar ploy to use discreetly posed women for attention. PETA in particular was blasted for trivializing animal suffering, projecting a marginalized image of activism, and concerning themselves with press coverage rather than tangible action. Taken together with their vegetarianism and the proclivity to work undercover while revealing unseemly practices in farms and laboratories, PETA, Lynx, and other animal rights organizations were (and still are) considered by some folks to be unnecessarily radical.[75] But for many folks the advertisements rang true, and the celebrity-laden, controversial media crusades reached a wide enough audience to become iconic. The result was that more people thought twice about wearing fur than ever before in American history.

Animal rights demonstrators were not the only influence on the precipitous decline of the fur industry in the 1980s. Market saturation was probably also a factor, as fur farms in Europe caused an excess of supply by overproducing pelts for exportation. The high availability of furs from Europe contributed to over 20 percent of American women already owning a fur garment. Thus, people who wanted to wear fur were probably already doing so, and were not buying many new products. In addition, messages from animal rights organizations that denounced the fur industry apparently failed to influence the public to the same degree with regard to leather, which was quickly replacing fur as a chic fashion.[76]

Once the stock market crash of 1987 plunged the United States into recession, the fur industry's deterioration was nearly complete. Only a fleeting demand from a strengthening Russian market would increase fur prices during the late 1990s, and this trend dissipated after the Russian economy collapsed in September 1998. Unrest in the Far Eastern countries that dictated demand from Asia also hurt the fur industry, and involvement of the United States in the Gulf War distributed resources away from resolving the recession.[77] One attempt at generating tax dollars—imposing a luxury tax on big-ticket items like jewelry, expensive cars, boats, aircraft, and, of course, furs—made most middle-class Americans even more wary of purchasing fur products. This perfect storm of animal welfare awareness and economic or political unrest in key countries dropped fur sales worldwide. In the United Kingdom, for instance, sales of fur fell 75 percent between 1985 and 1990.[78]

As a product with a niche market, take and sales of nutria were hit particularly hard. Both reached all-time lows during the fur market decline of the 1980s and '90s. Pelt values plummeted to $2.69 in 1988 and stayed low except for the short blip that was courtesy of the upwardly fluctuating Russian fur market. The nutria harvest, which was still a healthy 1.2 million in 1984–85, plunged to 114,646 in 1988—the lowest since the mid-1950s, and the first time in three decades that a yearly nutria harvest did not exceed 500,000.[79] Continued animal rights efforts, which were sponsored by celebrities like Pamela Anderson, Charlize Theron, Alicia Silverstone, and the Kardashian sisters, reiterated the case for making fur-wearing taboo. And with little market incentive for trappers to take nutria, harvest numbers continued downward throughout the 1990s.[80] Higher than the 15 million nutria in the 1970s and the almost 9 million in the 1980s, the decadal harvest of nutria dropped to 2.17 million in the 1990s—a low number considering the ballooning nutria population. Sales of trapping licenses in Louisiana diminished steadily and finally reached the "less than 1,000" milestone during the 2000–2001 season. By this time, there was virtually zero demand for nutria fur or any other pelts, and the fur industry never regained its former glory.[81]

The impact of the fur market decline was felt all over Louisiana both economically and environmentally. Reports of nutria damage to the wetlands were reaching the LDWF from coastal land managers in the years leading up to the three-decade low of nutria harvest in 1988, as trapping activity dramatically decreased. The damage was confirmed with aerial flights by LDWF. Armed with this evidence, a Nutria and Muskrat Management Symposium (NMMS) that brought together government biologists, private land managers, and other experts was organized. In 1992, symposium members concluded that the eating habits of nutria were adversely affecting wetlands. These findings were enough to convince the Barataria-Terrebonne National Estuary Program (BTNEP), to financially support more detailed aerial wetland damage surveys.[82]

Wetland damage from nutria was not entirely new. It began in the 1950s, after nutria multiplied when they were released from fur farms. But until the surge in anti-fur sentiment of the 1980s, the primary issue resulting

TABLE 4.1. Number of nutria damage sites and damage acres documented by the Nutria Harvest and Wetland Demonstration Program (NHWDP) and the Coastwide Nutria Control Program (CNCP) for 1998–2014. Reprinted with permission from the Louisiana Coastwide Nutria Control Program (CNCP) and the Louisiana Department of Wildlife and Fisheries (LDWF).

| Date | Total # of Damage Sites | Total # of Damage Acres |
|------|------------------------|-------------------------|
| 1998 | 170 | 23,960 |
| 1999 | 150 | 27,356 |
| 2000 | 132 | 25,939 |
| 2001 | 124 | 22,139 |
| 2002 | 94 | 21,185 |
| 2003 | 84 | 21,888 |
| 2004 | 69 | 16,906 |
| 2005 | 49 | 14,260 |
| 2006 | 31 | 12,315 |
| 2007 | 23 | 9,244 |
| 2008 | 23 | 6,171 |
| 2009 | 19 | 5,422 |
| 2010 | 11 | 2,260 |
| 2011 | 10 | 1,679 |
| 2012 | 11 | 1,129 |
| 2013 | 12 | 1,233 |
| 2014 | 11 | 1,115 |

from free-ranging nutria was damage to agricultural fields, not wetlands. Once additional aerial flights were funded and conducted, starting in 1993 and continuing until 2001, it became clear that wildlife managers needed to shift their focus to wetland damage, rather than sugarcane or rice field destruction.[83] Data from the surveys indicated that a whopping 4,726 acres of marsh were converted to open water, and at least 105,000 acres of wetland were damaged, often severely, from nutria.[84] Not surprisingly, the destruction was most salient in the parishes of Jefferson, Plaquemines, Terrebonne, and Lafourche, which were all significant players in harvesting nutria pelts for decades—the latter pair was usually ranked first and second, statewide, in pelt production.[85] Nutria were devouring Louisiana.

Some called it an ultimate irony—others, a vicious cycle. Historical desire for a flourishing fur industry was resulting in environmental havoc brought on by the descendants of escaped or released nutria. But without a healthy nutria harvest, wetlands would be turned into open water. And there was no sign of increasing harvests in sight. Whereas during the Great Muskrat Eat-Out of 1945–47, trappers were still working within a prosperous worldwide fur market, trappers were left with little incentive to harvest nutria after the anti-fur movement. The question of how to eliminate problem nutria became a hard one to answer.

But an answer was needed, and fast. Just as in the 1950s and '60s, when sugar interests were suffering because of thriving nutria populations, economic problems were imminent. The marred wetlands were the source of the Gulf Coast's seafood and tourism industries, and were the buffer between low-lying residential areas and hurricanes. Furthermore, range expansions of the Louisiana populations were taking nutria into Mississippi, and a once-local problem was becoming regional. *Mycastor coypus* was graduating from an introduced to an invasive species, and it was time for some damage control.

# Chapter 5

# Damage Control

In Louisiana, I don't think there is a serious predator on the adult nutria . . . of course, the alligator population is rather low so the effect on the nutria population is practically negligible.
—**V. T. Harris,** "The Nutria as a Wild Fur Mammal in Louisiana," Twenty-First North American Wildlife Conference (1956)

There are so many factors contributing to wetland loss, and nutria are definitely a part of that equation. They're not a native species, and we already have their ecological equivalent in muskrats . . . so a lot of people feel like if we could just press a button to get rid of all the nutria in Louisiana, we would do it.
—**Jacoby Carter,** research ecologist for the US Geological Survey, National Wetlands Research Center (2015)

*On a humid autumn morning in south New Orleans, five "zoo nutria" stand on the banks of a human-made lake in a walled enclosure. A sign saying "Nutria: The Mouse Beaver" adorns the makeshift habitat, telling the story of the inhabitants simply: "The scientific name of nutria* (Myocastor coypus) *comes from the Greek* myo *(mouse) and* castor *(beaver); coypu is the name for this species in its native South America. Nutria were introduced into Louisiana from Argentina in 1938 [sic]. As a result of escapes and later releases, they became widespread across south Louisiana in a mere five years, eventually spreading across the state and causing the decline of the native muskrat."*

*Nutria make excellent zoo animals, and Audubon Zoo staff maintain their nutria enclosure dutifully to educate the public about the dangers of invasive*

Going for a swim. Two nutria in an exhibit at the Audubon Zoo enter their human-made pond for a swim. The exhibit teaches visitors about nutria, their effect on the environment, and the dangers of invasive species. Photo by Theodore G. Manno.

*species—this sign notwithstanding. Perhaps left unchanged despite Martha Carr's 2002 examination of the H. Conrad Brote files at the University of New Orleans, and the author's 2015 reverification that Brote's SS* Del Norte *journals record "3 Cages 18 Live Nutria . . . On Deck" from an entry marked June 8–August 30, 1933, the "Mouse Beaver" sign shows the date of Ned McIlhenny's nutria purchase from the aforementioned "A[be] Bernstein," not the date of Chief Mate Brote's voyage five years earlier.*

*Of course, I am the only visitor who is not oblivious to the error. In fact, most visitors seem indifferent to the sign and hostile to the animals found within the enclosure. "It's a swamp rat," says one father to his sons, as they make a crude motion toward the nutria as if to shoot them with an imaginary rifle. As another family cruises by, a young boy looks at the nutria and imitates their apparent vocal repertoire. "Quack, quack," he says enthusiastically and confidently while making the requisite faux-shooting motion. "Is there any meat on them?" asks another young passerby to his mother. "Nah, they're just rats," says the mother as she cruises past the sign while cradling a cell phone to her ear, unconcerned with the formalities of technical biological classifications.*

*The indifference of the public seems reciprocated by the nutria, who care nothing of the visitors, the sign, or the environmental damage their species has caused. Wandering aimlessly about the enclosure, two nutria eventually slip into the pond for a swim; another pair walks the pond's perimeter, meeting the swimmers as they emerge from the water. The foursome walks to the back wall, and one sniffs another for no apparent reason. It seems like a poorly received sniff. "Quaaa! Quaaaaaa!" says the sniff's recipient, "Quaaaaa!" Soon, both sniffer and sniffee are making piglike grunts. "Quaaa! Snort! Beshhhhhh. . . ."*

*Meanwhile, another nutria is not interacting with the others as he hides under some vegetation. With nothing better to do, I talk to the solitary nutria about what is going on, how people are mad at them because of the damage they are causing. But the rodent refuses to acknowledge me, maintaining an empty look on its face. "This is what we do," the nutria seems to be thinking. "We didn't ask to be brought to a foreign continent. Now we're here, and we're going to eat everything because we don't care and we're nutria and that's what we do." I begin walking to the exit a few minutes later, and I hear a snort behind me—a piglike grunt from the lone nutria as I leave. As I depart, I wonder what the sound means and reflect on how nutria would be likeable and interesting animals if they were not eating the wetlands—at least I think so. Most nonbiologist visitors to the zoo probably disagree with me, recognizing nutria as a mere pest rather than wildlife. But there is one special day per year that comes every February, when rodents like nutria are celebrated instead of reviled.*

❖ ❖ ❖

Move over Punxsutawney Phil: T-Boy the nutria is the official weather prognosticator of New Orleans. Picked from the swamp-exhibit nutria at the Audubon Zoo each year, T-Boy makes a prediction that is a Cajun-style commemoration of Groundhog Day. The event is rooted in the German custom of Candlemas Day, where the weather on February 2 indicates when spring is coming. Every year, T-Boy emerges from whatever Louisiana-themed setup is constructed for that year's ceremony, such as the Superdome's field, a Mardi Gras float, or an Uptown Carnival parade route mockup. The latter was 2015's concoction—T-Boy surveyed the rutted streets and construction barricades, predicting a problem-free Mardi Gras and a bountiful crawfish season when he saw no shadow.[1]

Another official weather prognosticator is New Iberia native Pierre C. Shadeaux, a nutria named to reflect Louisiana's French influence despite the rodent's evolutionary origination from a Spanish-speaking country. Every year, a large crowd gathers just a few dozen miles from Avery Island at Bouligny Plaza in New Iberia to see Shadeaux and his custom-built, Acadianstyle residence. According to Mr. Shadeaux, who did not see his shadow this year, southern Louisiana should be ready for a long, lovely spring instead of a hot segue to summer.[2]

And living in the west New Orleans suburb of Metairie is a famous swamp rat named Boudreaux D. Nutria. He is a mascot of the New Orleans Zephyrs minor league baseball team. The other mascot is his wife Clotile. Together, Boudreaux and Clotile are King and Queen of Zephyr Field and ambassadors of baseball for southern Louisiana. Boudreaux or "Boo" has called Zephyr Field home since 1997, when he was found lost in the swamp exhibit at the Audubon Zoo. The Zephyrs adopted Boo and put him to work as the official mascot, and he married Clotile five days after a marriage proposal in front of a sold-out crowd on April 11, 1998. The nutria family now includes kits, complete with Cajun-sounding names—Beauregard, Cherie, Claudette, Jean-Pierre, Noelle, and Thibodaux. It was, as the Zephyrs website says, "love at first bite."[3]

All of these celebrity nutria are examples of one way to deal with the tragic loss of Louisiana's wetlands—have fun with the quirky, orangetoothed critter behind the damage. After all, nutria are entrenched in Gulf Coast folklore and part of Louisiana's charisma, even if they are destroying the shoreline. In a region that has experienced hurricanes, an oil spill, and economic strife during the past decade, folks can probably use the laugh. It is actually kind of a shame—if nutria were not eating wetlands, they might be very popular.

## Wasted Wetlands

Following the unusually active tropical cyclone season in 2005 and the Deepwater Horizon oil spill (also known as the British Petroleum oil spill, Gulf of Mexico oil spill, or Macondo blowout) in 2010, rapid loss of Gulf Coast wetlands became an issue of national concern. Most attention fell squarely on Louisiana because the state contains at least 40 percent of the nation's wetlands. Louisiana is also home to 15 percent of the freshwater wetlands in the United States and the Mississippi River alluvial plain, which is one of the 10 largest wetlands in the world. Maintaining the ecological health of these areas is a top priority for natural resource managers because wetland viability is tightly intertwined with the socioeconomic well-being of the southern coastal states. But unfortunately, a 1990 study indicates that

Louisiana is suffering 80 percent of the nation's wetland loss.[4] From 1978 to 1990, wetland loss in Louisiana averaged 34.9 mi.$^2$/year (90.4 km$^2$/year), a rate of about 0.004 mi.$^2$/day (1 hectare/day), with high pulses of forfeiture during hurricanes.[5] Almost 2,000 mi.$^2$ (5,180 km$^2$) of Louisiana's wetlands have become open water since 1932, and another 700 mi.$^2$ (1,813 km$^2$) are projected to disappear by 2050.[6]

Wetland loss causes many negative effects. Perhaps the most striking issue is that Louisiana's wetlands are the basis for over 55,000 jobs and billions of dollars in revenues. A significant amount of this economic activity comes from a $300 million commercial seafood industry, which garners more than 70 percent of its catch from species like shrimp, oysters, and blue crabs that count on coastal wetlands as a nursery for their young. Wetlands are also necessary for the seafood industry because they provide shelter for aquatic organisms and cycle nutrients from detritus and water. The latter promotes microorganism growth at the bottom of food chains.[7]

Another negative effect of wetland loss is the lack of protection from storms. Along with barrier islands, coastal wetlands are one of the most natural and least expensive buffers between storm surges and residential areas. With the potential to lower the speed or height of waves and floodwaters, wetlands are an environmental cushion between sea and land. In fact, wetlands can reduce water inundation on land by 5–40 percent, depending on the speed and strength of the storm.[8]

Wetland disappearance is also causing concern regarding the Gulf Coast's water supply, for three reasons. First, wetlands contain "biofilter" plants that shield us from contaminants in wastewater such as heavy metals, pesticides, and industrial or mining discharges. Second, biochemical processes that are crucial for life take place in wetlands. Examples include decomposition of organic matter by microorganisms, the nitrogen cycle returning nitrogen gas to the atmosphere, and reduction of harmful bacteria and viruses via filtration and absorption. Third, wetlands contain porous sediments that allow water to filter down through soil and overlying rock into aquifers that supply us with drinking water.[9]

Along the Gulf Coast, the past few decades have brought rising sea levels and subsiding coastal lands.[10] Effects of global warming on wetlands are also worrisome, because water from melted ice caps might increase sea levels even further. Another issue is that tropical storms and hurricanes may become more intense if they form over unusually warm waters. While some folks believe this conclusion is premature, many scientists subscribe to this notion and point to trends in meteorological data that are consistent with more frequent and intense tropical cyclones.[11] The loss of wetlands as a layer of storm surge protection would compound these storms, and surges are increasingly problematic even without the alleged influences of global warning on tropical cyclones. One mathematical model suggests that if wetlands were allowed to deteriorate with no restorative efforts during the next 50 years, storm surge heights would increase by 10–15 percent along Louisiana coastal areas to the east of New Orleans and result in massive destruction of property.[12]

Economic decline from wetland loss is inevitable. One study placed the economic value of wetland losses at $5.9 to $24.3 billion, with estimates ranging from $8,437 to $15,763 per acre and an annual loss of 25,500–27,500 acres per year. According to the author, these estimates were conservative because they did not include the costs of dismantling coastal infrastructure. Those consequences were valued at $756 to $815 million.[13] An earlier study estimated the annual value of wetland use on a per-acre basis as $37.46 for commercial fishing and trapping, $6.00 for recreation, and $0.44 for storm protection.[14]

Invasive nutria are one of the major reasons that Gulf Coast wetlands are disappearing. An LDWF study in 1999 found that over 100,000 acres of coastal marsh were damaged by hungry nutria.[15] Nutria herbivory (plant consumption) severely reduces the number and diversity of wetland plants and can cause the conversion of wetlands to open water. And unlike other common wetland disturbances like tropical storms or fire, which occur just a few times per year, nutria feeding is a year-round disturbance. Appropriately, *M. coypus* is listed by the Invasive Species Specialist Group (ISSG) as one of the top 100 "worst invasive species in the world."[16]

The mechanism of wetland damage from nutria is straightforward. When nutria are at a high density, they reduce vegetation biomass and create areas of open water called *eat-outs* that are vulnerable to further environmental disturbance. Nutria prefer the lower and more succulent portions of plants, so they usually feed by cutting vegetation near the waterline before swimming to a feeding platform. Thus, nutria eat-outs contain marshland that is denuded of vegetation, and remaining wetland plants are probably very sensitive to flooding or salinity.

Roots damaged by nutria feeding can no longer hold soil in place. Thus, sediment layers become increasingly sensitive to erosion, and organic matter that is normally used for replacing and recycling sediment is reduced.[17] This renders coastal marshes vulnerable to storm surges, which may extend further inland than they would without nutria grazing. Nutria may also compromise flood protection by digging burrows near levees. For example, Terrebonne port officials recently battled approximately 1,000 nutria that took up residence in and around the port's hurricane-protection levee.[18] Another phenomenon caused by nutria-induced soil instability is called *saltwater intrusion*, where saltwater infringes on areas that are normally freshwater. This profound habitat change likely impacts the distribution and abundance of various wetland species, particularly muskrats that rely on *Scirpus americanus* grasses in brackish marsh and resident or overwintering birds.

In contrast to the trends exhibited within the introduced range of nutria, a study of nutria density and distribution in the Pampas region of Argentina found almost no evidence of nutria-induced crop damage and concluded that nutria may not be a substantial threat to agriculture in their natural range. Surprisingly, the authors found that nutria density correlated positively with the availability of grasslands used for cattle grazing.[19] With so many reports of extensive damage from nutria along the Gulf Coast, these results beg the question of why the same trends do not apply to South America, where nutria reside as native wildlife.

We know how nutria were introduced to North America and how they spread across the continent. But why are nutria doing so much damage to American wetlands when they are relatively benign in their natural range? Three factors seem to explain this trend. First, nutria populations might be at a low density in South America if they are hunted heavily. Even when worldwide fur prices dropped, the economic worth of nutria in Argentina was still relatively higher when compared with the United States, owing to currency value and greater dependence on income-based hunting and trapping in rural areas. Population densities of nutria in South America are not known officially, but perhaps this dynamic increases the harvest of nutria beyond the level exhibited in the United States. Nutria harvesting methods probably also factor into damage-related statistics. Whereas South Americans tend to hunt or trap with more traditional methods, folks in North America often use costly motorboats and equipment. Thus, the return on harvesting nutria south of the equator is much more than what can be gained in the United States. Indeed, the expenses associated with harvesting nutria from remote areas is probably one of the reasons that American trap rates dropped so much in the 1980s and '90s.

The third and most compelling reason is that nutria are not native to the United States, and were casually placed into an already tenuous situation that was created over centuries by misguided anthropogenic activities. For example, researchers are examining the possibility that nutria proliferated wildly because they have few natural predators in American wetlands. Adult alligators eat nutria and used to be plentiful in Louisiana's swamps, but hunting decimated their populations so much that they were eventually afforded government protection under the Endangered Species Act (ESA) of 1973.[20]

The extent to which alligators might have controlled nutria populations is not eminently clear, but it is obvious that the introduction of nutria was an additional blow to the wetlands. With so many other issues that have accelerated wetland loss—levee building that prevents delivery of freshwater to marshes, canals that alter the area's hydrology, and now the recent hurricanes and oil spill—adding nutria to the equation was just about the worst thing we could do to an increasingly fragile and altered ecosystem.

## Part of the Equation

So how did humankind compromise Gulf Coast ecosystems and possibly make them more susceptible to the negative effects of nutria feeding? And how did we begin a cascade of wetland loss that was eventually continued by the presence of nutria? The answer lies in the flat landscape of southern Louisiana and how it results from years of natural geologic processes. Specifically, sediments that form Louisiana's coast eroded from uplands in the Rocky or Appalachian Mountains and were carried south by the Mississippi River. Plains around the Mississippi River originated during the end of the last ice age approximately 110,000–12,000 years ago, but the "youngest" parts of Louisiana were built into the Gulf of Mexico during the past 5,000–7,000 years.[21]

During this period, the modern Mississippi River delta resulted from a dynamic process known as the *delta cycle*, where the extreme lower Mississippi River snakes back and forth from east to west across the northern Gulf of Mexico, switching deltas every 1,000–2,000 years. Each delta cycle results in a new delta "lobe," which is created as the river obtains sediments from the continent's interior and deposits them near its mouth on the Gulf of Mexico. When enough sediment accumulates underwater, plants begin to grow and additional sediment accrues. Eventually, the delta "switches" because the river's path to the Gulf becomes longer and more difficult as the lobe builds. When the river inevitably responds by changing course, it abandons the older lobe, cuts a shorter route to the Gulf, and begins recycling the process. At least five major deltas resulting from this cycling can be distinguished along the Louisiana coast, and the oldest sediments are now furthest to the west.[22]

The constant ebb and flow of deltaic cycling creates a dynamic mosaic of habitats. Even elevation increases of a few centimeters from sediment accumulation can dictate post-storm flooding duration and which plant communities take root in the area. For example, marshes form where flooding creates water that is too high for trees to tolerate, and areas that are less waterlogged usually become forested. Deltaic cycling is also important because accumulating sediment brings nutrients that enhance marshland productivity.[23]

Deltaic cycling continued naturally when the active Mississippi delta was west of its current location. As a result, the Chenier plain was formed in the area west of Vermilion Bay from river sediments during two periods that geologists call the Teche Delta Period (ca. 2500 BCE) and the Lafourche Delta Period (ca. 1300 CE). Sediments accumulated there, and the mighty Mississippi River switched its course to find a route to the Gulf further east, forming the Saint Bernard delta. Meanwhile, abandoned lobes near the Chenier plain gradually sank and eroded, forming ecologically productive estuaries and creating barrier islands. Additional lobes formed with the river's new route, building stable land for taller vegetation. This process created a Mississippi River delta plain that stretched across 7,000 square miles at the time of European settlement, making it one of the largest river deltas in the world.[24]

Artificial cessation of natural deltaic cycling has drastically changed the area's environmental conditions since then. When Jean-Baptiste Le Moyne de Bienville established "Nouvelle Orleans" in 1718, the Mississippi River was active in the Belize delta lobe, now called the Birdsfoot delta. This lobe lies near the continental shelf's deep water and provided opportunities for waterborne commerce and transportation, so it was integral in the emergence of New Orleans as a shipping port. But by the mid-twentieth century, the Mississippi River was trying to take a more direct path to the Gulf of Mexico, and the Atchafalaya River, a distributary more than 100 miles west of the Birdsfoot delta, was capturing more of the river's flow. Since New Orleans and Baton Rouge would be finished as port cities if the river were allowed to complete its change of course, the US Army Corps of Engineers (USACE) installed a series of large water-control structures called the Old River Control Structure (ORCS) complex to artificially freeze natural cycling of the delta. Thus, the Mississippi River was prevented from its natural change in course.[25]

With a natural cycle, silt and clay sediments from the Mississippi River would deposit against the shoreline and eventually create mudflats covered with salt-tolerant vegetation. Instead, the ORCS preserved modern-day New Orleans and destroyed the sediment-building and land-creating processes that come with a river's shift in course. Even during the past 700 years, considered by geologists as the Modern Delta Period, very little sediment has entered the Gulf currents. The only exception seems to be Atchafalaya Bay, where 30 percent of the Mississippi River system's flow is diverted to the Atchafalaya River, and sediments carried by the river are extending the delta into the bay.[26]

Another reason for the deltaic plain's compromised ecosystem is the extensive levee system that was built for flood protection. Levees were constructed throughout the late 1800s and into the mid-1900s, including the East and West Atchafalaya flood-protection levees (EWAFPL), which were completed around 1940 and are more than 120 miles in length. As their names suggest, the EWAFPLs were on either side of the Atchafalaya Basin rather than along the Mississippi River, but they were nevertheless meant to protect against events like the Mississippi River floods that spilled into the Atchafalaya Basin in 1927.[27]

The side effects of these levees were similar to those from the ORCS. Not only did the levees prevent accrual of new sediment, but they also minimized the freshwater and nutrients flowing into important lakes like Maurepas and Pontchartrain. And although they were breached just a few years after they were built, levee construction continued steadily. With crevasses increasing in frequency as the levees grew in length and height, the negative environmental effects continued.[28]

Perhaps the most profound human-induced disruptions to the Gulf Coast ecosystem stem from the railroad that came in the 1850s. Few railroads passed through Louisiana prior to the Civil War. But that changed in 1852, when construction started on a railroad linking New Orleans and Jackson, Mississippi. The connection was via Pass Manchac, which resulted from a branch of the Mississippi River that carried sediment and built a narrow land bridge that separated modern-day Lake Maurepas and Lake Pontchartrain.[29] Opened within two years and rebuilt in 1865 after the Civil War, the railroad sprouted farming towns in areas along the tracks that used to be sparsely populated. Residents built levees to protect their farmlands from flooding and constructed houses on eight-foot-tall pilings.[30]

Although some farming towns along the railroad's route were annihilated by an unnamed hurricane in September 1915, the area continued to flourish as the tracks became part of a basis for industrial-scale logging. Large, inexpensive tracts of clear and open timberland were available, and demand for lumber was high in the Midwest and Northeast where timber resources were exhausted. With a steady flow of out-of-state capital, lumber emerged as the most central industry in Louisiana's economy, despite the continued importance of 'rat trapping and sugarcane. Prior to about 1880,

Wood eternal. An educational sign along a path at the Audubon Zoo with information on Louisiana's logging industry. Photo by Theodore G. Manno.

timber production in Louisiana was restricted to local needs and confined to areas near waterways. But by 1900, Louisiana ranked tenth in the nation for logging, and led the country in some years during the 1910s.[31]

During the high times of Louisiana logging, tycoons like Leonard Strader, Charles Hackley, Thomas Hume, and Joseph Rathborne acquired land at 25–50 cents per acre and systematically extracted massive amounts of lumber. Their enterprises became even more profitable after the Timber Act of 1876 allowed logging practices to become mechanized. For example, lumberjacks began using pull boats that enacted cables to drag fallen trees toward open water.[32] In the fledgling days of the logging industry, most ingathered trees were southern pine, especially longleaf pine (*Pinus palustris*). But cypress (family Cupressacae), referred to by some folks as *wood eternal*, eventually became king in the southern Louisiana swamps.[33]

With the largest cypress inventory in the United States, Louisiana usually led the country in cypress production during 1900–1920. The "cut out and get out" policy yielded tremendous short-term profits, and millions of acres containing 150- to 200-year-old virgin timber strands were cleared, especially in the western parishes. An area the size of New Jersey was clear-cut, and when the tree supply was exhausted, most big mills ran out of timber and closed down during the early to mid-1920s. Cypress mills were less likely to fold than pine facilities because Louisiana was more forested than swampland, but the effects of a wearied lumber stock were nevertheless felt across the region.[34]

Some consequences of unsustainable logging were environmental rather than economic. For example, aggressive lumber practices probably contributed to the disappearance of resident fauna. Three well-known species of now-extinct bird—the ivory-billed woodpecker (*Campephilus principalis*), Carolina parakeet (*Conuropsis carolinensis*), and passenger pigeon (*Ectopistes migratorius*)—inhabited the leveled swamps.[35] Another result of extensive logging was that canals and ditches that were built to facilitate transportation allowed saline water to move inland. This reduced plant biodiversity and shifted the habitat toward shrubby or scrubby vegetation as tall trees disappeared. Indeed, marshes tend to have fewer plant species than swamps, and studies indicate that some shrub-scrub plant communities are the result of logging activities and the changes they caused to salinity gradients.[36] Similar disruptions exacerbated the unnatural change in flora when canals built for the oil and gas industries allowed storms to shift seawater inland.[37]

Swamps with towering trees that were thousands of years old once dominated the Gulf Coast landscape. Following the disturbances caused by logging and levees, marsh now predominates the area. In addition, the succession that normally restores wetlands is probably altered permanently because environmental disruptions have all but ceased the naturally occurring fire that can be started by lightning (or occasionally by spontaneous combustion of dry fuel such as sawdust and leaves) that used to organically trigger succession. Prescribed burning is therefore used to increase wetland plant biodiversity, but the effects of invasive nutria juxtaposed over long-term hydrological changes continue to make wetland restoration challenging.[38]

The issues with attempting wetland restoration in habitat where nutria reside have been investigated with controlled experiments. When the interactive effects of soil nutrients, fire, and nutria herbivory are studied in the field, nutria herbivory emerges as a factor that reduces biomass and species richness when acting in concert with the other disruptions. For instance, results from a 1998 study suggested that *Schoenoplectus americanus*, a grassy flowering "sedge" plant that is a preferred food for muskrats and nutria, increased in relative abundance when mammalian herbivores like nutria were excluded.[39] When other researchers documented a shift in fauna dominance during a three-year study, the biomass of *S. americanus* increased

Floating muck. The floating muck of a complete eat-out. Courtesy of Louisiana State University Special Collections; T. O'Neil, *The Muskrat in the Louisiana Coastal Marshes: A Study of the Ecological, Geological, Biological, Tidal, and Climatic Factors Governing the Production and Management of the Muskrat Industry in Louisiana* (New Orleans: Louisiana Department of Wildlife and Fisheries, Fur and Game Division, 1949).

and dominant perennial grass *Spartina patens* decreased when protected from nutria. The researchers also indicated that the flora shift became more exaggerated and plant species richness declined when soil nutrients were altered. Thus, herbivory from nutria appeared to worsen the effects of reduced nutrient availability in soil.[40]

Until recently, nutria herbivory was one of the least quantified aspects of wetland loss.[41] Then, in 2010, the potential for nutria to intensify the negative effects of floods, levees, and other environmental disturbances like hurricanes was highlighted in a two-year study at Southeastern Louisiana University (SELU). The research involved vegetation fluctuations as a result of simultaneous changes in fertility and disturbance regimes, with the authors measuring species richness and total aboveground biomass after implementing a complex setup of five disturbance treatments (control, fire, herbivory, single vegetation removal, and double vegetation removal) and four fertility treatments (control, sediment addition, fertilizer addition, and sediment-plus-fertilizer addition). Trials included main plots that excluded herbivores like nutria, and variables in the experiment were selected with the idea of simulating the major historical factors affecting biomass production in coastal Louisiana.[42] Results suggested that nutria limited biomass production because the other disturbance treatments decreased biomass only minimally in the absence of herbivores. Nutria may have reduced the positive effects of fertilizer (a simulation of additional nutrients), which significantly increased biomass in experimental plots where nutria were not present. The SELU study also reiterated negative effects of saltwater intrusion caused in part by nutria feeding and other anthropogenic factors because the sediment-plus-fertilizer treatment, which simulated the additional nutrients and substrate material that freshwater would normally deliver, significantly increased biomass production.[43] Taken together with the studies on flora shifts, it seems that nutria presence may exacerbate the negative effects of levees, saltwater intrusion, and reduced accretion. As for direct nutria effects on the abundance or distribution of marsh animals, no reliable data from controlled experiments are available. But nutria herbivory impacts wetland plant communities, and probably affects animal species in wetland ecosystems.[44]

Does wetland loss from nutria feeding exacerbate the negative effects of hurricanes? The exact answer remains unclear. Hurricanes probably cause more economic and environmental harm if fewer wetlands are buffering the mainland from storm surge, and nutria presence is associated tightly with wetland forfeiture. But this conclusion is not shown directly with formal data, so the extent to which nutria may aggravate damage from hurricanes per se is unknown. Another factor to consider is that storms often drop nutria numbers substantially—dead nutria corpses are known to wash up on shores by the thousands after tropical cyclones. For this reason, wetland damage from nutria is probably not quite as immediate of a problem as it was during the 1980s and '90s. Nevertheless, it is troublesome that nutria seem to combine with other disturbances to accelerate wetland loss under experimental conditions; at a minimum, nutria are one of several values in the unfortunate equation for wetland disappearance. In light of recent events along the Gulf Coast, the possibility of nutria increasing effects from hurricanes warrants further study.

The same applies to nutria and their interactions with the effects of the Deepwater Horizon oil spill, which began pouring crude oil (a naturally occurring, unrefined petroleum product composed of hydrocarbon deposits and other organic materials) into the Gulf of Mexico by the hundreds of millions of gallons for 87 days starting on April 20, 2010. Geologists were already warning the public about Louisiana's loss of over 16 mi.$^2$ (41.4 km$^2$) of wetland area per year, in part because of nutria, and land forfeiture along the Louisiana coast was accelerated even further after the Deepwater Horizon oil rig exploded. With wetland vegetation dying soon after being coated with toxic oil, marshes and barrier islands that normally would have eroded in decades vanished in just a few years. One example among many comes from Barataria Bay, where a 12-mile-wide bowl of brackish waters and mangroves called Cat Island—a crucial 40-acre nesting ground for brown pelicans (*Pelecanus occidentalis*) and shorebirds (order Charadriiformes)— rotted into a streak of sand and a few dead branches after thick crude oil crept across the bay and swarmed the island's shores.[45]

The Gulf of Mexico's continued advance toward coastal communities increased the exposure of millions of residents to oil from the disaster. Were these negative effects of the Deepwater Horizon oil spill aggravated because eat-outs or other nutria damage permitted oil to penetrate deeper into coastal wetlands? As with questions regarding nutria interactions with hurricanes, there is not yet a precise answer. While the potential exists for wetland disappearance from nutria feeding to allow spilled oil to move further inland, no data from the recovery effort or experimental setups on this issue are available. The same applies to investigating whether nutria populations were negatively impacted by the oil spill, although almost no oil-covered nutria have come to shore and few nutria are seen in the salty marshes that are farther away from the coast than freshwater or brackish marshes, so contact of nutria with toxic crude oil from the spill may have been limited.[46] With most Deepwater Horizon oil spill–related research focusing on protected species, the possibility that nutria-induced wetland loss worsened the effects from the incident deserves more investigation.

## Later, Gator

When nutria were implicated in accelerating coastal wetland loss during the 1980s and '90s, the ethics of wearing fur was a topical issue. But during the past two decades, scientists have recognized the role that large-scale alligator hunting may have played in allowing nutria to reproduce and spread. Because alligators are generalist feeders and one of the apex predators of the wetlands, some scientists maintain that alligators could potentially control populations of nutria. Under this hypothesis, alligator presence could indirectly increase wetland biomass, shift species composition toward naturally occurring species, return land accretion and nutrients, and increase post-logging tree regeneration in marshes via feeding on nutria.[47]

American alligators almost met the same end as passenger pigeons and Carolina parakeets following the "cut out and get out" logging policy in Louisiana from 1900 to the 1920s. Alligators are the largest reptiles in North America, and at one time they were also one of the most plentiful. A popular anecdote showing alligator abundance in the region comes from naturalist William Bartram, who hobnobbed with Seminole chiefs and was the first white man to penetrate the dense tropical forests of Florida. Bartram endured various encounters with the aggressive reptiles during his expeditions in the Southeast. In one of his memoirs from 1791, he wrote:

> Should I say, that the river . . . from shore to shore, and perhaps near half a mile above and below me, appeared to be one solid bank of fish, of various kinds, pushing through this narrow pass of St. Juans into the little lake, on their return down the river, and that the alligators were in such incredible number, and so close together from shore to shore, that it would have been easy to have walked across on their heads, had the animals been harmless.[48]

Another oft-mentioned quote comes from John James Audubon, who witnessed alligators on the Red River. According to Audubon:

> They were so extremely abundant that to see hundreds at a sight along the shores, or on the immense rafts of floating or stranded timber, was quite a common occurrence, the smaller on the backs of the larger. . . . The shores are yet trampled by them in such a matter that their large tracks are seen as plentiful as those of sheep in a fold.[49]

Reports like these started to appear less frequently as the 1800s passed because alligators were harvested for boots and saddles, and oil produced from alligator corpses was used in steam engines and cotton gins. Use of alligator hides continued to increase during the Civil War, as development of commercial tanning processes in New England led to production of more durable hides.[50]

Demand for alligator products fluctuated during the next 150 years. Meanwhile, free-ranging alligators failed to replace themselves quickly. Alligators may live up to 70 years and can gain about a foot in length per year. Adults reaching about six to eight feet become breeders, but mortality among young alligators is high—only 10–20 percent survive long enough to reproduce, and many young are cannibalized by larger individuals.[51] While Audubon recognized alligator abundance in his memoirs, he was also keenly aware of these reproductive limitations and the imminent decline of alligator populations from hunting pressures: "It was on that river [the Red River] particularly that thousands of the largest size were killed, when the mania of having either shoes, boots, or saddle-seats, made of their hides, lasted."[52]

Ned McIlhenny was also aware of declining alligator populations. Widely considered an expert on the species, Ned's monitoring of an enclosure that he filled with the giant lizards while running the Tabasco operations resulted in a seminal book on alligators. Titled *The Alligator's Life History* and published in 1935, Ned McIlhenny's report was that alligators "fairly swarmed" prior to widespread hunting.[53]

Ironically, it was soon after McIlhenny's celebrated account that alligators started to approach endangerment. When interest in alligator products began in the 1800s, only alligators more than 2.4 meters (almost 8 feet) long were harvested. By the 1930s, every alligator that could be captured was killed, regardless of its length. While the exact historical population of alligators in coastal Louisiana is unrecorded, anecdotal evidence suggests that by the 1950s, an increasingly cavalier attitude toward alligator hunting led to overharvesting and resulted in a miniscule and unsustainable alligator population. Thus, state officials suspended alligator hunting in 1962. As they worked to eliminate poaching, the US Fish and Wildlife Service (USFWS) listed *A. mississippiensis* as endangered in 1967.[54]

Muskrat population increases during the 1940s culminated in the Great Muskrat Eat-Out. But before muskrats spread westward from the Mississippi delta in the late 1800s, they were rare in Louisiana's coastal marshes.[55] Interestingly, nutria populations also skyrocketed during the 1940s and '50s. And like muskrat increases, the rise in nutria numbers coincides with the years when alligators declined. Does this mean that alligator hunting caused the increase in nutria and muskrat populations? As with the questions regarding nutria, hurricanes, and oil spills, the answer remains unclear.

Because alligators eat nutria and muskrats, trappers used to express concern that they might be competing with alligators.[56] On the other hand, nutria are so prolific that some folks think alligators are unlikely to have a significant impact on nutria populations. So how much do alligators eat nutria (and muskrats)? Two studies from the early 1940s report that muskrats were found in 33–52 percent of examined alligator stomachs. In 1961, a dietary survey of 25 alligators found nutria in 56 percent of the stomachs examined, with nutria comprising 46 percent of the stomach volume in these alligators.[57] Occurrence of nutria in alligator stomachs dropped to 5 percent the following year, but a 1987 study found that muskrats and nutria combined accounted for 83 percent of alligator diet weight and occurred in

77 percent of the examined alligator stomachs. Muskrats occurred in these stomachs with greater frequency than nutria, but more than two-thirds of the mammalian flesh weight in the alligator stomach samples was from nutria. Thus, nutria seem to be commonly predated by alligators, although nutria were found only in the stomachs of alligators over three feet in length.[58] These results lend some circumstantial support for the notion that alligator hunting is at least partially to blame for nutria and muskrat population increases.

Nevertheless, it is sometimes unclear as to whether nutria population reductions are driven by alligator feeding, and conclusions regarding alligator consumption of nutria are often compounded by other factors. For example, a study on the Sabine National Wildlife Refuge (SNWR) seems to show that nutria density dictates alligator diet. Population estimates and fur harvest records for the SNWR indicated that nutria numbers were declining rapidly during 1961–65, from an estimated population of around 74,000 to under 10,000. During periods when nutria populations were close to 74,000 individuals, nutria remains were identified in 56 percent of examined alligator stomachs. In contrast, when nutria populations were under 10,000 individuals, less than 7 percent of alligator stomachs contained nutria.

If alligators were causing fewer nutria in the refuge, then why didn't more alligator stomachs contain nutria remains? According to the authors, the reason for the decline of alligator predation on nutria was probably the nutria control program that was taking place during that period, which accounted for a 30 percent reduction in the nutria population. Thus, alligators were probably eating nutria more if they were available, rather than causing a decline in nutria per se. The authors understood that no data existed to conclusively show that alligator predation was causing fewer nutria at the SNWR—only a casual relationship between alligator and nutria numbers was established.[59]

A recent population modeling study of alligator-nutria interactions addresses the debate regarding how much alligators influence nutria populations. According to the mathematical simulations, the presence of large alligators (but not small alligators, which are less likely to eat nutria, especially larger females), negatively impacts nutria populations. The authors also suggest that controlling the harvest of large alligators may have a significant impact on nutria populations and serve as a means to help control nutria impacts to wetlands.[60] Because the study addresses a topical debate in wetland ecology, it has garnered attention as progress toward defining a relationship between alligators and nutria. But the study is also noteworthy because its angle provides an interesting way to think about biogeochemical processes in wetlands.

The novel approach comes from a "top-down" view of marsh restoration processes, which focuses on how predators like alligators might control the abundance of species that are lower in the food chain.[61] Before this study, wetland restoration approaches were typically "bottom up" and focused on how systems were regulated by nutrient availability from sediment. This approach also assumed that all organisms live under conditions where they are competing over a shortage of resources. Instead of maintaining the classic bottom-up philosophy, the authors propose a "trophic-cascade" hypothesis based on their simulations, in which energy through the system is dictated by apex predators at the top of the food chain, rather than limiting nutrients at the bottom. Indeed, the simulations indicated that hunting reduced the density and average size of alligators, and this reduced the effectiveness of alligators as a natural control on fast-populating wetland herbivores like nutria.[62]

The trophic-cascade hypothesis is challenging to test with field data because systematic data on nutria population numbers, marsh integrity, plant biomass, and alligator densities must be collected across multiple sites. With no empirical data from controlled experiments, it is impossible to propose an exact number of alligators that is needed to control a nutria population. Lack of experimental data also leads critics of the hypothesis to focus on instances where strong indications of top-down control on nutria are lacking. For example, a 1994 study showed that Terrebonne Parish outranked all other coastal parishes in alligator nesting density, but was also ranked first in nutria herbivory damage.[63]

Other anecdotal evidence showing low nutria abundance where alligators are numerous supports the trophic-cascade hypothesis. The freshwater maidencane-dominated marshes around Lake Boeuf, an area that is largely free of nutria damage despite a mean density of 81.8 alligators per mile during 1978–98, is one example. Another is the Jean Lafitte National Park (JLNP), where the alligator population seems fairly high and features old, large individuals because alligator harvest is not allowed.[64]

The simulations, taken together with these anecdotes, support the idea that alligator presence can indirectly increase plant biomass, land accretion, and tree-regeneration rates via predation on nutria. Experimentally derived data indicating a direct cause-and-effect relationship between alligators and nutria are unavailable, but perhaps future researchers will provide more systematic evidence that supports the trophic-cascade model. Meanwhile, wetland resource managers are shifting toward a holistic approach of

monitoring species interrelationships and away from the traditional practice of managing a single species.

Thus, officials are recognizing the importance of alligators to wetland
ecosystems and are working to preserve their interactions with other species. These efforts have been largely successful; populations started to
rebound after the LDWF began monitoring alligators in 1970. Populations
recovered so well that an alligator-harvesting season was reinstated in 1972,
beginning with Cameron Parish and eventually becoming statewide in 1981.
Despite the harvest, alligator nest numbers increased by an average of 13
percent each year from 1970 to 1993.[65]

Louisiana state officials also began an alligator-farming program in 1986
by issuing permits to collect wild alligator eggs for hatching under captive,
artificial conditions.[66] The farming program, along with ESA protections
for wild alligators, culminated in removal of A. mississippiensis from the
USFWS endangered species list in 1987. With a population of nearly 2 million in about 4.5 million acres of mostly freshwater marsh habitat, Louisiana
now has more alligators than any other state.[67] Alligators also fared well
nationwide following ESA protections, as alligators currently range across
the southeastern United States from Texas eastward to North Carolina.[68]

One challenge brought on by this approach is finding a happy medium
between conserving alligators to provide a natural control of nutria and
harvesting alligators to encourage economic activity. For now, populations
have rebounded so well that alligators are routinely killed for food, clothing,
or to prevent conflicts with humans. For example, a 30-day annual harvest
season for alligators takes place in Louisiana each fall, and an experimental "bonus tag" program in which trappers are issued 10 percent more tags
than they would normally receive was initiated in 1999. During springs and
summers on Marsh Island, harvest has been allowed during predetermined
periods. The intention is to harvest smaller and more frequently occurring
four- to five-foot-long alligators, which are less likely to consume nutria
than full-sized six- to seven-foot-long alligators, so this program may repre-
sent a compromise between alligator harvest and natural control of nutria.[69]

If nothing else, these popular programs are generating income that can
be used for other conservation and management initiatives. Using 2001
as an example, 34,583 wild alligators that were usually five to seven feet in
length were harvested for their hides and meat to the tune of over $9 mil-
lion.[70] How much these alligators could have reduced nutria-related damage to wetlands remains unclear. But with no hard data to definitively say
how many alligators will control a certain amount of nutria, most managers
feel it is probably prudent to continue these programs.

Alligators are worth millions of dollars as a novelty for swamp-tour companies in addition to their potential value as nutria control. This remains a consideration for resource managers who are determining how much to harvest alligators.[71] Meanwhile, research supports the need for comprehensive management approaches that improve wetland ecosystems as a whole and manage the interrelationships between nutria and their potential predators, such as bald eagles, great horned owls, red foxes, great blue herons, red-shouldered hawks, and raccoons. Careful management of coyotes (*Canis latrans*), a nutria predator with populations along the Gulf Coast that may be growing, could also prove to be important in the future. However, it is also apparent that prudent management of potential nutria predators is unlikely to emerge as a comprehensive solution to invasive nutria populations. Perhaps LDWF biologist Edmond Mouton provided the best articulation of this reality during an interview in 2011: "While all of these [possible predators] might eat nutria, they are obviously not controlling the population."[72]

## Restoration Efforts

Some wetland sites that are experiencing damage from nutria will not recover and are gradually being converted to open water. But a majority of damaged sites are probably capable of recovering and can undergo restoration. Without efforts to restore wetlands consumed by nutria or compromised by other factors, coastal communities may continue to disappear and more folks will leave the Gulf Coast region.

So how are resource managers restoring the areas that are damaged by hungry nutria and muskrats? Following the 1990 passage of the Coastal Wetlands Planning, Protection and Restoration Act (CWPPRA), there are myriad ongoing restoration activities in the Gulf Coast region. Designed to fund construction of coastal wetland restoration projects throughout its 20-year lifespan, the CWPPRA has used an approximately $30–80 million per year budget to start 151 authorized coastal restoration or protection projects since its inception. Over 110,000 acres of Louisiana wetlands have benefited from CWPPRA initiatives.[73] Primary restoration strategies for wetlands in the Mississippi delta are focused on diversions of water across levees to augment inputs of freshwater, nutrients, and sediments to wetlands, and reducing use of wetland areas for development or logging.[74]

Freshwater diversions are a key component of wetland restoration because the current levee system interferes with water flow. Diverting freshwater toward swamps also brings sediment and nutrients into the deltaic wetlands and may restore original flood levels. Examples of diversion plans in action include the Hope Canal and the Bonne Carré spillway west of New Orleans. Another example comes from the Lake Pontchartrain Basin Foundation's (LPBF's) recommendations in their 2005 conservation management plan for four sub-basins in the Lake Maurepas and Lake Pontchartrain regions. The report suggests diverting freshwater from the Mississippi River to the area and using treated sewage to enhance marsh production.[75] So that this nutrient-containing water may flow freely into the plains—a process scientists call *sheet flow*—some researchers suggest that existing canals should be plugged and levees gapped or breached with openings.[76] Limiting saltwater intrusion from storms by closing outlets with engineering additions may also halt salinity corruption in freshwater marshes.[77]

Reducing urban sprawl is another part of wetland restoration strategies.[78] Constructing homes and businesses in wetland areas is anathema to freshwater diversion efforts, since resultant floods may place structures at risk. But asking landowners to allow flooding on their property as part of reestablishing sheet flow is unlikely to be met with a positive response. Thus, building further on areas of damaged wetlands has been discouraged in favor of acquiring land for the public trust. Some folks suggest that large tracts of land remaining from former logging leases may provide an opportunity for public maintenance of resources.[79]

Meanwhile, some private land is still subject to logging. Termination of this logging, which is primarily for cypress, would likely help with wetland restoration efforts. But logging continues because a multitude of bureaucratic issues have occurred. For example, the US Army Corps of Engineers (USACE) was able to restrict logging near waterways until a US Senate committee drafted the Water Resources Development Act (WRDA) of 2005. The WRDA contained a provision to remove this power from the USACE, even though the act focused primarily on providing Louisiana with funding for coastal restoration. The Louisiana House of Representatives voted 92–6 in support of the USACE-related provision, and the Louisiana senate approved it unanimously, even though the provision was apparently added without public notice or discussion.[80]

Nutria are thwarting some of the efforts to recover land that was damaged by their feeding behavior. A prime example is the challenge brought on by nutria gnawing at baldcypress trees that are planted to restore disturbed land to previous conditions. While nutria do not seem to cause lethal damage to mature trees, their gnawing may make baldcypress more susceptible

Dead cypress. A dead cypress swamp north of Falgout Canal in Lafourche Parish, Louisiana, photographed during 1998. Courtesy of Louisiana State University Special Collections: George Castille Slides, Louisiana Digital Library, Baton Rouge, Louisiana.

to other stressors. Nutria also pull at young seedlings, eat succulent bark from the taproot, or clip seedlings aboveground.[81]

Whether nutria substantially impact long-term natural regeneration of cypress swamps is unknown. Either way, some reseeding efforts have been suspended because of nutria damage. Occasionally, nutria have caused complete failure of planted or naturally regenerated tree stands, and a variety of repellents and barriers have been used to combat the issue. Using wire-mesh fencings or hardware-cloth tubes around stand perimeters appears to be one effective technique for keeping hungry nutria away from seedlings. But these materials are expensive, labor intensive, and difficult to use. Plastic-mesh fencings can be cheaper and easier to use, but they are usually ineffective because nutria can chew through them.[82] As a result, the issues of protecting seedlings from nutria continue to frustrate land managers.

Results from a recent study in Oregon on Vexar plastic-mesh tubes seem promising. Researchers found that woody plants protected by Vexar tubes demonstrated 100 percent survival over a three-month initial establishment period, while only 17 percent of unprotected plantings survived. The study also reported that nutria were more active in unprotected areas when compared with protected areas. Thus, plastic-mesh tubing may emerge as an effective strategy for mitigating nutria damage to seedlings.[83] Or, if no physical protection from nutria is available, seedlings may be protected with the more subtle method of planting in the autumn rather than the spring. Research indicates that baldcypress seedlings, even when unprotected, suffer less damage when planted on this timetable because other food sources are available to nutria in October and November.[84]

Vegetative biomass decreases in response to nutria herbivory, and baldcypress is not the only flora affected. *Sagittaria latifolia, S. platyphylla, Spartina patens, S. alterniflora*, and many other species of marsh vegetation are often lost to hungry nutria and are hard to recover with nutria present.[85] One way to combat nutria interference with revegetative wetland areas is to plant functionally equivalent plant species that are less preferred by nutria and other herbivores. For example, red osier dogwood (*Cornus sericea*) and willow (*Salix* spp.) can be planted instead of black cottonwood (*Populus balsamifera* ssp. *trichocarpa*) trees.[86]

If nutria are interfering with restoration efforts, then they probably need to be controlled. This realization has led state officials to address a fundamental issue—too many nutria—while conducting other restoration projects. But by the time folks started brainstorming ways to rid the wetlands of nutria, hunting and trapping were no longer lucrative because the international fur industry was in economic free-fall and public disfavor. Still, the nutria needed to be "ratted out." So, in an ironic twist, officials made a calculated decision to revisit the same actions that created the fur-for-all atmosphere in which nutria were originally transported to Louisiana. They brought the trappers back to the swamps, and made sure it was worth their while.

# Chapter 6

# Bounty Hunters

The influence of nutria alone is sufficient to cause the marshes to continually decline, jeopardizing their existence.

—**Allan Ensminger** in J. D. Addison's *Nutria: Destroying Marshes the Old Fashioned Way*

*It's a fashion show like no other. On a winter night in 2010, more than 20 designers, including Cree McCree, née Mary Jane but called by her partial surname since college, are gathering to exhibit nutria-themed fur products in the House of Yes!, a funky Brooklyn art gallery. Sweet tea flows, musicians play swamp rock, and people are merry.*

*The spectacle follows three similar efforts in New Orleans and a Nutria-Palooza! On the Bayou event 140 miles away from the coast in Lafayette, where models donned fur coats, hats, or jewelry made from nutria as they sauntered down the catwalk—or in this case, the ratwalk. Folks dined on nutria stew, and a silent auction closed out the 'Palooza!, with attendees purchasing nutria-trimmed Mardi Gras dresses and nutria hats—what Jerry Seinfeld's ex-girlfriend Elaine Benes would call "rat hats."*

*Michael Massimi of the Barataria-Terrebonne National Estuary Program (BTNEP) gave the keynote speech that night in Lafayette, discussing swamp rats, the damage they inflict on wetlands, and the need to restore Louisiana's coastline. Massimi was an appropriate choice. The BTNEP, an organization based in Thibodaux, awarded a grant to self-described "Bohemian artist–type" Cree McCree and made the 'Palooza! happen.*

*Cree's clothing company, fittingly called Righteous Fur, promotes and sells fur fashions made from nutria. A portion of all proceeds goes to the BTNEP.*

*From her abode in New Orleans, McCree tells me that the purposes of the 'Palooza! and other nutria-clothing events are to raise awareness about the unique environmental issues on Louisiana's coast, to help control an invasive species, and to defend wetlands via resource utilization.[1] Of particular concern to McCree is the nutria bounty program run by LDWF, which provides financial incentive to hunters and trappers for traveling to state-sponsored locations and redeeming tails taken from nutria they dispatch. About 97 percent of the bounty collectors do not use meat or fur from harvested nutria. As Cree explains, this means that a few hundred thousand nutria corpses are buried underground or tossed into a watery grave each year.*

*McCree argues that because nutria are being killed as a pest—at one point, Jefferson Parish's SWAT team used nutria for target practice, and now nutria are killed via the bounty program—then they might as well be utilized. The LDWF also promotes this philosophy, but not overtly with fashion shows featuring nutria on the runway—they simply offer information on buyers and dealers of nutria products. Either way, the idea of using nutria for meat or fur has not caught on with most of the public, much to the chagrin of McCree. "I think we should be looking to reuse them, in fashion and food," says an enthusiastic, chatty McCree as she describes her "Cree-ations." "It seems like a criminal waste for them just to be chucked back. The decaying bodies pollute the wetlands too.*

*"If you look at how hunting and fishing started, with Native American traditions, there's a culture of honoring the animal and using every part of it. Instead of honoring that practice, we're just cutting off the tails and throwing away the corpses so they can rot."*

*When McCree came to Louisiana from New York in the late 1980s, she did not know about "swamp rats" and the damage they were starting to cause. Her fashion shows promoted using natural and recycled materials, until a friend told her to pop some nutria into the mix. Two BTNEP grants later, McCree says that she's not an invasive species person per se, but rather a person who is trying to raise awareness of the wetland loss along the Gulf Coast.*

*"They're having problems in places like Maryland and Oregon too, but so much is at stake here in Louisiana where we need every acre of wetland that we can have, and where other stuff [like hurricanes and oil spills] is going on. So much else is happening environmentally and [tending to the nutria invasion] is just one thing that needs to be done."*

*McCree's philosophy is not without its critics. Some folks worry that nutria fur might catch on, or least gather a concentrated following, and that this could lead to the massacre of other innocent furbearing animals. After all, when someone wears fur on the street, the public does not know from what animal it came, and the fashion may become* en vogue. *Thus, opponents of McCree's "pest fur" idea point to massive and sudden historical swings regarding what is popular, accepted, believed, sought after, and commercialized, and hope the fur fashion shows will not catch on. A similar argument surrounds the carving of fish bones, bovine bones, and even fossilized mammoth ivory for jewelry as an alternative to elephant ivory—the fear is wearing real bone of any kind could make folks crave elephant ivory again.*

*But McCree says that although the vast majority of bounty collectors do not use the meat or fur from harvested nutria, she has not seen any overt resistance to her creative usage of dead nutria. Strangely, almost no organized opposition has come from animal rights activists who would "rather go naked than wear fur," as some advertisements used to proclaim.*

*"I have some vegans and vegetarians designing. There will always be some people that are not into using the fur, but people [in Louisiana] have a good understanding of what's going on environmentally. Remember, people here go out and shoot nutria and have been living like that for generations. It's not New York City. People understand that it's a necessary evil."*

*McCree is also more open-minded than most Americans regarding human consumption of nutria. "I think that eventually the whole idea of eating nutria meat will not be all that strange," says McCree. "I've had nutria meatballs and they taste good when they're really spicy. It just tastes like meat with spice . . . but it's not like chicken, and not a real gamey flavor either.*

*"And even if you don't eat them, it's beautiful fur . . . so why not make something beautiful?"*

✦ ✦ ✦

In 2006, a then-obscure defender for the National Football League's New Orleans Saints named Stephen Gleason led his team to victory against their archrivals, the Atlanta Falcons, by blocking a punt during the Saints' first home game in the Superdome following Hurricane Katrina. After the Superdome's roof was damaged while thousands of people endured Katrina inside under deplorable conditions, the crucial play became a symbol of recovery in New Orleans. The legend of Gleason's defensed kick developed further when the Saints won their first Super Bowl in 2009 behind the quarterbacking of Drew Brees, who was rejuvenated from a serious shoulder injury that many thought would end his career. Brees became another icon of resilience as New Orleans rebuilt, but sadly, Gleason did not play for the team that won the Super Bowl. Two years after his famous blocked punt, Gleason revealed that he was battling Lou Gehrig's disease. He was awarded a symbolic championship ring by the Saints, and in July 2012, a statue depicting Gleason's play entitled *Rebirth* was raised outside the Superdome.

After Gleason—who is apparently unrelated to the soldier named Gleason who may have provided Edmund McIlhenny with his peppers—was diagnosed, he collaborated with a filmmaker to document his life for his unborn son. Included in the footage was a 12-minute audio clip taken in a hotel conference room before a 2011 playoff game between the Saints and the San Francisco 49ers, during which New Orleans defensive coach Gregg Williams allegedly encouraged his players to injure the opposition. The audio, which was not authorized by Gleason for release, nevertheless became public on April 4, 2012, and was part of a notorious "bounty" scandal that led to suspensions of Saints players and coaches by the NFL for using money to reward hits on members of competing teams.[2]

Despite this scandal, people in New Orleans remember the names of Brees, Gleason, and other players during the post-Katrina era as a source of cultural pride, and Brees's black-and-gold #9 jersey is worn all over town. Thus, many Saints fans were angry with the NFL for handicapping their beloved Saints. The team was part of life in Louisiana, just like shrimping, crawfish, gumbo, café au lait with beignets, Bourbon Street, antebellum homes, and, to many folks, hunting or trapping.

Katrina (in 2005) combined with other hurricanes in 2008–9 to wipe out thousands of nutria. But nutria were still damaging wetlands, which was the last thing Louisianans needed following the storms. The problem first became intolerable in the 1980s, when the market for nutria pelts declined. By the 1990s, nutria were eating a substantial amount of wetland vegetation in several states, particularly Louisiana. With over 105,000 acres of nutria-related damage found by survey flights that occurred throughout the 1990s, LDWF mounted a response. It was clear that even though wetland restoration techniques would be paramount in recovering the Gulf Coast's post-nutria ecosystem, they would not be a comprehensive solution. Removing one of the sources of the wetland damage—the nutria—was also a necessity to ensure the long-term health of wetland areas.

So, while the Saints were undermanned following the "bounty" scandal, another type of bounty was incentivizing nutria harvest. The rationale for this enticement followed the same philosophy that dictated LDWF programs for decades—place value on a resource, have hunters and trappers make a living harvesting it, and allow the process to control overpopulation. Government agencies discussed many ideas for ousting nutria from Louisiana, but the most practicable and immediately effective method involved using financial incentive to return hunters and trappers to the wetlands.

This decision came on the heels of similar programs that were instituted in other nutria-plagued areas. Some were implemented with the intent of eradication, and others were merely hoping to contain nutria populations that could not be completely stopped from mushrooming. For the most part, bounty-incentivized hunting and trapping caught on with the public. But as far as using nutria for meat and fur—not so much. Either way, the irony is palpable. Most people think that our best chance to rid the environment of a nuisance animal whose presence can be blamed largely on creating stock for hunting and trapping is to revive hunting and trapping. And the question of what to do with the corpses of harvested nutria still remains.

## On the Ratwalk?

The Coastwide Nutria Control Program (CNCP) is a well-oiled operation that takes place in 19 coastal parishes in the area south of I-10 and I-12. Once funding was secured from the Coastal Wetlands Planning, Protection and Restoration Act (CWPPRA) and the Coastal Protection and Restoration Authority (CPRA), the LDWF began paying a bounty on nutria tails starting with the 2002–3 season. The payment per tail was originally four dollars, and increased to five dollars in 2006–7 to encourage continued participation.

A few hundred individuals participate each year, and LDWF reviews their registration vigilantly. Each participant must officially register, obtain a valid license, and provide documentation of the landowner's permission for them to kill nutria on the site if applicable. To provide financial incentive for harvesting nutria, LDWF has established six collection stations in easily accessible public places. These locations are where nutria tails are counted, participants are presented with a receipt or voucher to collect a bounty, and records concerning the number and locations of collected tails are maintained. Between seasons, LDWF continues to survey for nutria-related damage to see how well their nutria-control efforts are progressing—counting the actual number of nutria is too herculean of a task.[3]

The CNCP was created to reduce wetland damage attributable to nutria and to regulate nutria numbers. Eradication per se, which is considered unrealistic, has never been an objective. This conservative approach is appropriate because the areas covered by nutria are vast, and some are bound to be inaccessible to hunters or trappers. Indeed, LDWF biologists have estimated from nutria harvest data that there are still several million nutria in the Louisiana coastal zone. Thus, most folks use the "you can't stop them, you can only hope to contain them" qualification when evaluating the CNCP, and in these terms, the program is generally regarded as successful. Almost two-thirds of the nearly 400 CNCP registrants claimed at least one bounty during the 2013–14 season, and this resulted in the submission of 388,264 nutria tails worth $1,941,320 in incentive payments—an average of about 20,000 nutria taken from the swamps per week.[4]

During the 2013–14 season, some registrants did not claim any incentives and more than one-quarter of participants turned in less than 200 tails. But one-fifth of CNCP participants turned in between 200–500 tails, one-eighth submitted between 500–800 tails, and over 40 percent of the participants redeemed 800 or more tails. This means that almost half of CNCP participants earned a take-home pay of several thousand dollars, and the high level of take has translated into reduced damage to coastal wetlands. According to the CNCP, 1,115 acres were damaged by nutria following a survey in 2014, which is down from 1,233 in 2013 and a fraction of the 14,260 acres damaged during 2005, the year of Hurricane Katrina.[5] Although this reduction will not return the thousands of acres that were lost to hungry nutria during the past half century, these data nevertheless show a favorable trend toward limiting further wetland loss.

Hurricanes have been the only factor to slow the CNCP. The reason is that when coastal residents must rebuild infrastructure, they become less focused on hunting and fishing than in years without storm encounters. Indeed, the smallest number of nutria harvested during the CNCP was 168,843 for the 2005–6 season, directly following the impact of two major hurricanes (Katrina and Rita), and totals for most other years have been over 300,000. But the decrease in nutria harvest following hurricanes has been mitigated by other environmental factors that increase participation in the CNCP. For instance, nutria harvests tend to increase slightly when water levels are high because this allows hunters and trappers to penetrate deeper into the marsh and concentrates nutria into areas of high ground where they are more easily hunted. Not surprisingly, the 2009–10 season featured a period with very high water levels and generated a take of 445,963 nutria, the highest yearly nutria harvest during the CNCP.[6]

Despite the zealous efforts to eliminate nutria, LDWF remains concerned about harvesting nutria ethically. About three-quarters of harvested nutria are taken via shotgun rather than trapping, and hunters are not allowed to use lead in their shotguns, lest the other wildlife that might feed on dead nutria be poisoned. Proper disposal of the tails that are redeemed for bounties is also monitored, as all tails must be delivered to an approved disposal facility.[7]

As for the tailless corpses, CNCP participants are supplied with a buyer and dealer list to encourage harvesting nutria for fur and meat. Some trappers skin nutria themselves to sell the meat and fur, with one trapper making an extra $7,000 for his efforts.[8] But with a miniscule market for nutria products, the vast majority of CNCP participants do not absorb the labor and logistical costs of utilizing the entire corpse. For instance, only around 3 percent of the nutria harvested during the 2013–14 season were utilized for meat or fur (or both), down from a 3.3 percent utilization rate during the prior season. The remaining corpses were discarded via underground burial, abandonment in heavy overhead vegetation, or placement in a body of water.[9]

## Other Management Practices

LDWF accepted incentive-based hunting and trapping as the most appropriate course of action after studies indicated that wetland loss from nutria feeding would be mitigated and that direct nutria control would be supported by the public. Other responses to nutria feeding were researched, but none were considered to be as immediately effective as the bounty program that is now operating. However, some of these other methods may help reduce nutria damage and can be used in concert with hunting and trapping to create a comprehensive approach for controlling nutria populations and preventing wetland loss.

### Fences and Barriers

The shortcomings of using barriers to reduce nutria damage include high costs for materials and other "non-target" animals becoming excluded along with the nutria. Protecting a large agricultural field or wetland area with fencing, for instance, is usually not practical because a very large amount of fencing is needed, and it is not ecologically advantageous to exclude animals besides nutria from a wetland. However, fences are common and effective measures for barring nutria from gardens and lawns.

To prevent nutria from gnawing at trees, hardware-cloth tubes or wire-mesh fencing around the perimeter of the tree is usually effective, although extensive use can become impractical and expensive. Sheet-metal shields and plastic seeding protectors can sometimes prevent nutria damage, but neither is used widely. The former are regarded by many as costly eyesores, and nutria can chew through the latter.[10]

## Water and Slope Management

Nutria construct dens by burrowing into embankments. When water levels rise above the dens, nutria will burrow farther into the embankment or dig new burrows that are closer to the water's surface. To prevent erosion from fresh diggings, water managers discourage this behavior by keeping water levels steady.

A trial-and-error process often begins if steady water levels do not stop nutria from digging. During the winter, managers may drive nutria away from an area by raising water to almost flood level, which forces the rodents out of their dens and into the hard-to-tolerate cold weather. Over the summer, managers may also try extreme drops in water levels to expose nutria dens to predators, which cause nutria to search for a more protected area.

Another option is to eliminate all standing water from a drainage site or embankment, since nutria often find any area that holds water to be attractive. This technique is almost impossible to implement in low-lying wetland areas and lacks formal data to support its effectiveness. But some land managers find the practice to be helpful for their particular situation, particularly if the issue involves damage to agricultural ground.[11]

Proper land-management and farming techniques can indirectly curb nutria-caused damage. For example, nutria are liable to burrow into steep slopes that are covered with vegetation, so ridding the area of dense vegetation, thicket, and weeds with mowing or cutting tends to minimize human-nutria conflict. If more radical measures are necessary, then land managers sometimes grade or contour bank slopes to make them gently sloping, which is an angle that is less favorable for nutria burrowing. When nutria remain persistent despite these measures, then bulldozing active burrows becomes an option.

Fences and barriers are often combined with water and slope-level management for maximum effectiveness. But even with careful land management or sweeping changes to a landscape, nutria are often unconcerned with slope angles, water levels, or other minutiae in their habitat and may become quite inventive when constructing burrows. When this occurs, land managers have no choice but to dispatch the offending nutria or relocate crops, fields, and gardens. New agricultural areas are planned away from waterways where nutria live, natural vegetation buffers are placed next to water to make crops and gardens less attractive to nutria, and livestock are kept away from nutria dens so that they do not break their legs in burrows.[12]

## Harassment and Repellents

No chemical repellents are registered for nutria, and using unregulated repellents is illegal. Most repellents available in the United States are for birds, and commonly used compounds such as methyl anthranilate and anthraquinone can effectively repel birds, but not rodents.[13]

Rodent repellents like Thiram may repel nutria, but their effectiveness remains unclear. Exploring this possibility is probably not worthwhile because of the time and expense needed for experimentation and the issues that are inherent in repellent use. The primary issue is that repellents merely shift target animals from one area to another, leaving damage to occur in another locale. Another problem lies in the delivery of repellents. To make a repellent stay on vegetation long enough to potentially affect nutria, adhesive would need to be applied as well. Given the vast areas that are inhabited by nutria, repeated applications would probably require millions of gallons of both the chemical and adhesive. Taken together, these issues mean that repellents are not regarded highly as a method of controlling damage from nutria.[14]

Like most wild animals, nutria will try to escape from a threatening situation. Folks have therefore attempted to harass nutria in order to keep them away from vegetation, crops, or human-made structures. Loud noises, high-pressure water sprays, and large barking dogs (small dogs can be intimidated by nutria) are the most common forms of nutria harassment. Other methods of pestering nutria that are subtler include stuffing their burrows with rocks, sprinkling the area with the urine of predators (usually mink, coyote, or bobcat), leaving cat litter nearby, or exposing their tunnels with shovels. Humans usually tire of these techniques quickly and nutria are resourceful enough to return shortly after harassment, so this type of nutria control is usually short-lived and minimally effective.[15]

## Toxicants and Fumigants

Zinc phosphide is the only rodenticide that is currently registered for nutria control. The chemical is a grayish-black powder with a garlic-like smell and is used widely for other rodents as well. When applied to fresh baits at a high concentration, zinc phosphide can kill almost all of the nutria in an area quickly. But zinc phosphide is almost never used for nutria control over a large area because it is extremely toxic for the environment. Zinc phosphide takes many months to degrade and leaches into soil; it also poisons animals other than nutria that may be threatened with extinction or otherwise crucial to the area's ecological balance.[16] Because of these issues, zinc phosphide is subject to strict federal and state regulations that govern the concentration applied to an environment, and its use is limited to certified pesticide applicators.[17]

Another issue with using zinc phosphide to control nutria is the expense and labor-intensive effort that is necessary to place the bait. Pre-baiting is required to incite nutria feeding in an area, and this necessity results in several visits to potential treatment sites before using any bait. Placing zinc phosphide–laced bait on the ground usually results in the poisoning of non-target animals, and baiting watery areas involves time-consuming construction of floating rafts that contain the rodenticide. Meanwhile, heavy rains and high humidity can react with bait and render it ineffective within weeks. Replacing the bait increases labor and adds expense, and so does the disposal of nutria carcasses that contain toxic material. Only about 25 percent of poisoned nutria die where they can be located, but they still need to be collected and either buried or burned to prevent exposure to other organisms.[18]

Although a few fumigants, such as aluminum phosphide, are registered for controlling burrowing rodents, none are registered for killing nutria. Some fumigants may develop as options for nutria control in the future. But for now, no scientific data exist regarding their effectiveness in controlling nutria or potential side effects to the environment.[19]

## Induced Infertility

Some readers may wonder why nutria are not simply given birth control. One method of contraception—removing or sterilizing all of the males in an area—is obviously impractical. Even if all males were removed or sterilized, invasive nutria usually exist in widespread and contiguous populations, so immigrant nutria would probably come and impregnate most of the females anyhow. In addition, notions about feeding nutria bait that is laced with a now-illegal synthetic estrogen called diethylstilbestrol (DES) are in disrepute because of secondary effects. Other animals eat the bait and are rendered unable to reproduce, as is the case for predators of nutria that wind up ingesting DES indirectly.[20]

Contraceptives such as chlorohydrin, mestranol, diazacon, and a vaccine against gonadotropin releasing hormone (which inhibits the production of sex hormones) are indicated only for other species, usually ground squirrels, mice, rats, or birds.[21] They have no known delivery method that is effective for widespread nutria populations. Given the number of nutria involved, delivery of contraceptives via dart gun or by capture-and-injection strategies is impracticable and cost-prohibitive. Most contraceptives also involve chronic exposures, which require repeated applications of bait that are usually executed by aircraft, and these exposures would occur over the entire environment and therefore affect key non-target animals like pelicans and muskrats, or maybe even economically crucial species like shrimp. No formal studies test the dosage of contraceptives needed for nutria or consider the potential environmental implications. Both are prerequisites for approval of the practice by the FDA. For these reasons, en masse sterilization of nutria is not being seriously considered as a method for eliminating problem nutria.

## Beyond the Third Coast

From the 1880s to the 1960s, nutria made their way not only into America's Gulf Coast region, but also into many other states and countries. The route of introduction usually involved escapes and releases from captive populations that were bred for their pelts. Occasionally, nutria appeared in an area because a population expanded its geographic range and spread from one state or country to another. Sometimes the introduced nutria multiplied and graduated to "invasive" status after they became free-ranging. In other instances, introduced nutria failed to establish a long-lasting population, especially if weather conditions in the area were cold or dry.

While the highest densities (and some of the most bizarre stories) of released nutria have usually occurred along America's Third Coast, other incidents of escape or release have enabled nutria to wreak significant havoc on wetlands or crops. Eradication or control efforts in these places have yielded varying levels of success. These notable regional case studies are worth examining to improve nutria and wetland management in the Gulf Coast and elsewhere because they offer insights regarding which practices worked, which ones failed, and how to prevent the environmental problems associated with misplaced nutria.

### Chesapeake Chew-Out

Data on the number of countrywide nutria sightings suggest that, during the 1980s and 1990s, more nutria resided in Maryland than in any other state besides Louisiana. The population was established when nutria escaped, or were released, from fur farms near the ecologically fragile Blackwater National Wildlife Refuge (BNWR) in Dorchester County. Humid wetlands bordering the Chesapeake Bay suited nutria well, and by the 1990s, up to 50,000 nutria were chomping away at the BNWR and other parts of the Delmarva Peninsula, a nearly 200-mile-long isthmus between the bay and the Atlantic Ocean.[22]

Nutria feeding resulted in erosion of marsh throughout the Chesapeake Watershed. Substantial portions of BNWR wetlands were eaten-out and turned into a vast lake clogged with silt, with more than 7,000 acres of marsh converted to open water. As a result, the BNWR was no longer supporting the diverse array of wildlife that always depended on the marsh. The USDA blamed nutria for the loss of nearly 50,000 acres of Chesapeake-area wetlands, and adverse consequences were felt all across the region. Some were quite costly or threatened human health and safety—increased flooding, filling of ditches, damage to corn and soybean crops, timber die-off as saltwater from the bay intruded farther inland, and the collapsing of dykes, roads, water-control structures, and stream banks. As was the case in Louisiana, nutria also competed for forage with native wildlife, and Maryland's muskrat numbers reduced significantly.[23]

Damage to the Chesapeake Watershed occurred despite efforts to control nutria shortly after their introduction. Hunting and trapping of nutria continued until the late 1970s, when nutria populations crashed during the severe winters of 1977–78. But nutria populations suffered few long-term effects from the cold weather, and they rebounded over the next few years. Unprecedented marsh loss soon followed, leading the BNWR to respond in 1989 with a "trapper rebate" program that paid trappers $1.50 for each nutria they killed. Federal lands were leased for trapping native muskrats, and the BNWR funded research projects to estimate nutria numbers. Meanwhile, the Maryland Department of Natural Resources (MDNR) and the US Fish and Wildlife Service (USFWS) established a multiagency task force—the first of its kind—to investigate how to combat the feral nutria populations.

The task force quickly determined that recent marsh loss was due primarily to nutria feeding and that damaged marshes would recover if nutria were removed. A meeting with representatives from federal, state, and private organizations like the Maryland Cooperative Fish and Wildlife Research Unit (MCFWRU), the MDNR, University of Maryland Eastern Shore (UMES), Tudor Farms (TF), and the USFWS called the Nutria Control Summit was convened in June 1997. One year later, a three-year, $3.8 million pilot plan to evaluate nutria eradication was implemented. As part of the plan, biologists determined strategies for trapping nutria while minimizing impacts to other species, and determined how many nutria would need harvesting in order to decrease populations.[24]

Because the Chesapeake Bay is the nation's largest estuarine ecosystem, the nutria issue there grew into an issue of federal consideration. Additional funding totaling $2.9 million was sought to achieve the pilot plan's objectives, and this money was eventually authorized by the secretary of the interior. Other grants were also acquired from the USDA and Congress over the next few years. With the pilot plan establishing eradication as a realistic possibility and delineating a plan for nutria harvesting, a full-fledged nutria-eradication phase was implemented in April 2002, with the USDA's Animal and Plant Health Inspection Services (APHIS) Wildlife Services assuming the primary project responsibilities. Shortly thereafter, President Bush signed the Nutria Eradication and Control Act of 2003, authorizing expenditure of $4 million per year over five years to eradicate nutria from the Chesapeake-Delaware Bay area. The extra support apparently came just in time. A 2004 study commissioned by Maryland state officials determined that nutria damage could potentially cost Maryland around $37 million each year in lost economic activity if nutria were left uncontrolled.[25]

The nutria-eradication project has been expensive, costing approximately $13.8 million as of 2011. But the efforts have also been indisputably successful. Taken via guns, traps, and sometimes with the assistance of dogs, thousands of nutria were eliminated during the mid-2000s. In addition, most landowners with nutria on their property were cooperative with APHIS staff and leaped at opportunities to eliminate the nutria as eradication efforts progressed. This allowed the project to expand into a few hundred thousand acres over the entire Delmarva Peninsula, leading to near-elimination of invasive nutria in the region. Eradication efforts are continuing on the Eastern Shore of the Chesapeake Bay, although with so many nutria already killed over the years, much of the project now consists of monitoring for remaining pockets of nutria that are deep in the swamps.[26]

Even though the APHIS-directed program was successful, nutria-eradication and wetland-restoration efforts on the Delmarva Peninsula continue to garner federal attention. President Barack Obama signed the Chesapeake Bay Protection and Restoration Executive Order (CBPREO) in 2009, declaring the Chesapeake Bay as a national treasure and requiring federal agencies to protect and restore the health, heritage, natural resources, and natural sustainability of its watershed. As per the order, the Department of the Interior is working with federal and state agencies in Maryland to expand public access to the Chesapeake Bay, restore an additional 30,000 acres of wetlands, and refurbish wetlands that have been degraded by nutria.[27]

While wetland-restoration efforts are under way, wildlife officials continue to fight the resilient nutria. Every so often, the appearance of nutria in an area that was considered to be nutria-free reminds the officials that their work, though efficacious, is not yet finished. For example, biologists recently discovered nutria along the Manokin River, where nutria have not been spotted for years. Nutria also unexpectedly turned up a few years ago along a 15- to 20-mile stretch of the Eastern Shore's Wicomico River.[28] Wildlife officials in Delaware continue to monitor situations like these closely, lest some wandering nutria cross into their state. Folks in Virginia are nervous too—nutria were once seen in a marsh near Virginia's sliver of the Delmarva Peninsula, and concern exists that some small nutria populations near Virginia Beach could migrate into ecologically sensitive areas.[29] Even officials in New Jersey are aware of the potential for nutria range expansion. As widespread efforts in the mid-Atlantic sought to eliminate nutria for good, 2007 brought a nutria sighting in southern New Jersey's wetlands along the Delaware Bay.[30] The current whereabouts of the rogue nutria, and whether he or she was traveling with friends or family, remain a mystery. Nevertheless, the possibility exists that nutria may have recently added New Jersey to their ever-expanding range.[31]

### Northwest Passage: The Gold Nuggets in Fur

Fur trappers and traders laid the Oregon Trail during 1811–40, and this tradition continued into the twentieth century when nutria farms were established in Oregon during the 1930s and '40s. Around 600 farms raised nutria, mostly in the Portland and Tillamook areas, and the descendants of released or escaped nutria spread to coastal and inland portions of the state.[32] Some of Oregon's feral nutria may also be descended from nutria that were released to control the state's Brazilian water weed (*Elodea densa*), which siphons oxygen from coastal lakes.[33]

The first recorded release of nutria in Oregon comes from a Tillamook County fur farmer in 1937, and other unrecorded releases apparently led to widespread feral nutria populations by 1946.[34] Meanwhile, trade in captive nutria continued and increased the potential for nutria to become wild. With local periodicals touting nutria as "gold nuggets in fur," at least three organizations, including the Oregon Purebred Nutria Associates (OPNA), Nutria Incorporated (NI), and the Purebred Nutria Association (PNA), promoted Oregon's nutria trade in the 1950s. State and county fairs also presented breeding stock while publicizing nutria as an option for get-rich-quick fortune seekers. Meanwhile, OPNA set up offices in Portland, Salem, and Grants Pass to handle sale and registration of their "purebred" animals to gullible buyers, advertising a hefty price of $950 per breeding pair and listing a ménage à trois of two females and a male for $1,550.[35]

As with Louisiana and Maryland, the nutria fad came and went quickly. Within a decade of OPNA selling threesomes of nutria for the price of a small car, the market for nutria pelts plummeted and nutria furs became worth less than a can of coffee. Oregon's nutria-raising farmers cut their losses and released their stock into the brush or neglected to maintain captivity pens as nutria tunneled their way to freedom. Limited interest in harvesting wild nutria developed, with nutria catch reports increasing sharply from 29 in 1957–58 to 5,950 in 1971–72. Large nutria pelts of good quality sold for up to seven dollars in the 1970s. But the 1980s anti-fur movement that ruined enthusiasm for trapping in Louisiana similarly affected the Northwest's fur industry, and Oregon-based nutria populations were left to proliferate unmolested.[36]

Numerous issues have resulted from the growing nutria populations in Oregon. Burrowing from these nutria damages crops, water-control devices, stream banks, field borders, and farm ponds. Native plants or agriculture that have been destroyed by nutria include alfalfa, grass seed, wheat, barley, oats, field corn, sweet corn, carrots, table beets, cauliflower, cucumbers, melons, sugar beets, and Oregon oak (*Quercus garryana*). One farmer hired a wrecker three times in two years to pull his tractor out of a field that was collapsed by nutria burrows; another lost a cow when it fell through a nutria burrow that was undermining the pasture. Ground destabilization caused by nutria digging is to blame for an Oregon farmer who was killed when his tractor toppled into a drainage ditch.[37] And the strangest nutria-related issues come from conflicts between family dogs and nutria in residential areas—a mostly irrelevant dynamic in Louisiana, where nutria usually live deep in the wetlands. One particularly sad case involved a family dog that was killed by a trap set for invasive nutria. In another troubling incident, a misguided nutria lover shot a neighbor's dog that came onto his property and chased the rodents that he apparently befriended by feeding them bread and leftovers.[38]

In 2001, the Center for Lakes and Reservoirs (CLR) at Portland State University was assigned to develop an Oregon Aquatic Species Management Plan (OASMP), including outreach, prevention, detection, research, and damage mitigation. The CLR produced a range map of nutria density in Oregon, and other researchers conducted an inconclusive enclosure-controlled herbivory study to analyze the effects of nutria feeding habits.[39] Meanwhile, nutria continue to be an occasional nuisance on private property. To combat this issue, Oregon has outlawed spreading nutria via relocation and classified nutria as "unprotected nongame wildlife." This means that nutria may be trapped or shot without a license.[40]

The neighboring state of Washington also endures environmental damage as a result of invasive nutria. Farms that dealt in nutria fur were established all over Washington starting in 1941, with locations in Seattle, Bothel, Maple Valley, Bellingham, Bremerton, and probably others. Nutria from escapes or releases out of these farms were discovered near Lake Washington and the Snohomish, Skykomish, and Snoqualmie Rivers later that decade. At some point, these populations probably spread into British Columbia.[41] The rodent quickly devoured native plants, including various species of willow (*Salix* spp.), cattail (*Thypha* spp.), and native rush (*Juncus* spp.). Vegetable gardens were no safer, as nutria exhibited a proclivity for cabbage.[42] Nutria also caused erosion and damage to embankments. Attempts by land managers to restore native habitat during the 1980s and '90s were often unsuccessful because nutria ate the new plants, with one project near Vancouver losing almost $400,000 for this reason.[43]

The Washington Department of Fish and Wildlife (WDFW) is encouraging the public to control nutria on a local, case-by-case basis. But no official eradication program exists, and landowners are responsible for the costs of nutria control. Among the extermination options recommended by WDFW are the USDA Wildlife Services, which eliminate nutria via trapping and shooting. Such efforts are particularly common in Skagit County, where some agricultural groups raise money to employ trappers for nutria control because nutria frequently damage embankments.[44]

## Other Domestic Cases

California conducted a small-scale, well-known, and successful eradication program that ousted *Myocastor coypus* from the state by 1978. The several unwanted nutria brought to Elizabeth Lake by William Frakes during his ill-advised foray into nutria farming probably did not result in the eradicated populations—Frakes's nutria were apparently washed into an isolated cave by a flood—although their fate has never been entirely clear. It is more likely that the eradicated populations resulted from subsequent importations of nutria to California that took place after Frakes's death in 1920 because California did not have a sizeable population of feral nutria until 1940.[45]

Probably because the dry conditions in Southern California are unfavorable for free-ranging nutria, *M. coypus* never flourished.

In contrast, the southeastern United States is generally hot and humid. Escaped or released nutria are more likely to flourish in this climate. Thus, the states of Florida, Alabama, Georgia, Virginia, Tennessee, Arkansas, and North Carolina (but apparently not South Carolina—yet) all have viable nutria populations resulting from escapes, releases, or imprudent translocations. Most have not resulted in widespread agricultural or ecological damage, but wildlife managers remain vigilant for nutria in areas where they are not normally seen.

The most worry comes from a booming nutria population in the watersheds of eastern North Carolina that has expanded into Virginia. Among other issues, nutria are burrowing between wastewater retention ponds and a nearby river, which could potentially cause drinking-supply contamination. Both Virginia and North Carolina allow lethal nutria control if they threaten structures, and informal efforts from the public have ousted some of the problem animals. But considering the resiliency of *M. coypus*, officials in both states are mapping reports of nutria and seeking assistance from the federal government to gear up for Maryland-style eradication efforts, should they become necessary.[46]

Florida was home to at least 20 nutria farmers that were raising 300–400 head of nutria in the 1950s. Several hundred feral nutria, resulting mostly from purposeful releases to control aquatic weeds, were reported to be roaming the state during this decade.[47] Apparent descendants of these nutria are sometimes reported in the Tampa Bay area, although the distribution of nutria in Florida is unknown and currently of minimal concern.[48] This is also the case for some other southern states where nutria were introduced (e.g., Arkansas, Tennessee, Georgia), although the possibility of a sudden population increase always looms.

Nutria died out in some states when they reached the wilderness. Either cold or dry weather did not provide nutria with a suitable habitat in which to live and breed, or the small number of nutria released into the wild was not enough to form a breeding population. All indications are that small populations of nutria that once existed in the colder-than-average mountain-west states of Utah, Idaho, Montana are now extinct; populations that apparently remain in Colorado and New Mexico are small and sparse. Wild nutria have also been reported in eight midwestern states but are probably now extinct in all of them. The only possible exception is Oklahoma, where there may be a miniscule nutria population remaining. In Indiana, feral

nutria were swiftly eradicated; other nutria introduced to the Midwest simply failed to survive and breed.[49]

## Nutria Worldwide: The British Invasion

Nutria came to the British Isles in 1929 when over 50 fur farms were established in the East Anglia region, which includes Sussex, Hampshire, Devon,
and Norfolk. Many of these farms were not properly maintained, and this
allowed some nutria to escape their captivity and form breeding populations in the countryside.[50]

At first, no one saw the British invasion coming. A few nutria escapes
were discussed blandly in 1935 when a researcher named T. Warwick fate-
fully wrote that "it does not seem as if the coypu would readily establish
itself in the first place, and its extermination, if necessary, should not prove
difficult."[51] Instead of satisfying Warwick's prediction, nutria increased their
range during the 1940s through the 1960s, and their numbers likely reached
around 200,000. Widespread damage ensued. Nutria devoured valuable
crops, especially sugar beets. Ecosystems with native wetland plants also
suffered, as nutria ate-out reed swamps and turned flowering rush (*Buto-
mus umbellatus*) and cowbane (*Cicuta virosa*) from relatively common
plants into rarities. In low-lying East Anglia, nutria burrowed into drainage
systems and caused extensive damage that put the region at an increased
risk for flooding.[52]

The public demanded governmental action in response to the nutria invasion. They received it in 1962 when the Ministry of Agriculture launched a
three-year hunting and trapping campaign to reduce nutria populations and
confine surviving nutria to a single area. Sweeping efforts resulted in a 90
percent population reduction, but this was a short-lived decrease. After an
unusually cold winter in 1962, biologists realized the strong possibility that
weather, not trapping, accounted for the smaller nutria populations. Indeed,
nutria numbers rebounded during the warmer winters that followed, and
a few thousand nutria were left in East Norfolk after the campaign. Despite
the continued harvesting efforts, nutria populations started doubling every
year. By 1975, the number of nutria in England totaled around 19,000.[53]

Upward population trends made wildlife managers concerned over
whether trapping would eventually eliminate the nutria. So scientists at the
Coypu Research Laboratory in Norwich, which was established to conduct
research for the control program, studied the viability of certain areas for
trapping and determined how many trappers and years would be necessary to accomplish complete nutria eradication. The result was an eventual

answer to East Anglia's nutria problem—a concerted, Agricultural Department-led 10-year eradication campaign starting in 1981 that employed 24 salaried trappers and enacted intensive trapping over the entire region. Eight years later, the trapping efforts were terminated when 21 months passed without a trapped nutria and with only two elderly males observed in the wild.[54] Other than a solitary incident in 2012, in which a "giant rat" that authorities believe may have been a nutria was killed in County Durham, *M. coypus* has not been seen in England since the eradication. Thus, the eradication program in Britain is widely regarded as one of the most successful anti-nutria initiatives ever.[55]

## Nutria in Mainland Europe

From 1882 to 1967, nutria were bred in captivity and turned loose in nearly every mainland European country. Exceptions include Albania, Luxembourg, Portugal, and half a dozen former Soviet republics, although an asterisk must be placed on these countries because no data exist on the historical presence of nutria. The most recent reports list nutria as extinct in Scandinavia and Ireland, although a single nutria that may have been someone's pet was recently spotted in the Irish countryside.[56] Feral populations of nutria still exist in the other mainland European countries where nutria were introduced, although in some of these countries, namely Bulgaria, Hungary, Romania, and Spain, nutria are probably the result of range expansions from neighboring countries.[57]

Although nutria have not caused widespread damage in Europe, they remain a source of frustration in parts of the continent where populations are particularly large. Perhaps the worst nutria-related damage is in Italy, where nutria were introduced as early as 1928. When nutria spread from the boot-shaped Italian mainland to Sicily and Sardinia, they devoured rice fields and interfered with efforts to restore eaten-out wetlands by chewing on planted seedlings.[58] These issues led to overt control programs in some regions. A 1995–2000 study estimated losses of €11,631,721 (over $12 million USD) in nutria-related damage, mostly to crops and riverbanks, and over 220,000 nutria were removed from Italy at a cost of €2,614,408 (almost $3 million USD).

But Italian nutria populations remain resilient, and some control programs have done little to reduce nutria numbers. One example among many is an intensive trapping campaign during 1994–96 in a northern Italy wetland that removed 8,600 nutria but failed to reduce the overall number of nutria in the area.[59] Some nutria-control efforts may lack effectiveness

Worldwide. Nutria have been introduced to continents worldwide, including Europe, Asia, and Africa. Courtesy of Bildagentur Zoonar/Shutterstock.

because they occur mostly in locations with high densities of nutria and after damage has already taken place.[60]

From 1974 to 1985, feral nutria in France increased their numbers so much that they were controlled with anticoagulant poisoning.[61] Nutria also devoured beet crops and damaged levees in low-lying regions of the Netherlands, necessitating European agencies to start a control-by-trapping program. Control efforts are decreasing damage but nutria populations endure, as thermal pollution (the harmful release of heated liquid into a body of water or heat released into the air as a waste product of industry) in certain rivers is allowing nutria survival despite the harsh Holland winters. Meanwhile, migrating nutria from Germany and Belgium are crossing into the Netherlands and creating woe among Dutch farmers. Whether northern European populations of nutria will one day require more comprehensive control remains to be seen.[62]

### Africa, the Middle East, and Asia

At some point before 1958, nutria were introduced to Zimbabwe, Zambia, and possibly Botswana.[63] A few reports from the 1990s place nutria in these countries, but they have apparently never become established in the wild.[64]

Nutria might become a nuisance in Kenya, where captive stock escaped in the 1960s after importation to the Kinangop Plateau for fur farming. The newly feral nutria settled near Lake Naivasha and pythons (*Python rebae*) were released as control, but to no avail. Nutria therefore remained in Kenya, and were officially reported there as recently as 2005.[65]

In the Middle East, nutria are present in Israel and Jordan. They escaped from fur farms during the 1940s through the 1960s, and they cause damage to fishponds and river systems.[66] Nutria are also present in Russia following the country's illustrious fur history, which includes a successful large-scale introduction of nutria for fur farming during 1926–34. Nutria that escaped or were released from these southern Russia operations acclimatized to local lakes and rivers, although cold weather probably limited nutria population growth. Nevertheless, nutria are iconic enough in the (former) USSR that a stamp printed circa 1980 honors their contribution to the fur industry.[67]

The former Soviet republics of central Asia are one of the few places where nutria are not considered pests. Some people there raise nutria in a "semicaptive" context, where habitat is managed such that nutria can live through winters. Ponds are drained and vegetative cover is provided for protection from the cold before nutria are purposely released and acclimatized to the habitat. Sometimes, nutria are even fed throughout the winter as if domesticated. Enthusiasm for nutria runs particularly high in Azerbaijan, where farmers raise selectively bred "white coypus."[68]

The "nutria itch" is not nearly as prevalent in East Asia, where nutria found themselves in China during efforts to start fur farms in the early 1960s. No official reports place feral nutria in China, but they are likely residing near the southern coast.[69] Chinese nutria populations spread when Thailand imported nutria from China in 1993; five years later, nutria were reported to be roaming the jungles. While nutria have not become so destructive as to demand an organized control program, the Thai government is monitoring the situation and has prohibited importation of additional nutria. Meanwhile, some of the feral nutria have been eaten by villagers.[70] Reports of nutria sightings have also come from India, and whether these nutria result from trade with China is unclear.[71]

The most salient nutria problem in Southeast Asia belongs to Japan, where nutria were introduced in 1910 by military officials who encouraged nutria farming because they believed the pelts to be of high quality. When the international fur market declined, Japanese nutria pelt prices plummeted and nutria stock was often killed or released into the wild. These introductions have necessitated attempted eradication of nutria via hunting since 1963.[72] The Japanese conflict with nutria has even been illuminated in a tanka-form poem:

*Shirasagi mo Sugamo mo Koi mo kechirashite,*
*Sasagase gawa wo Nutoria yuku.*
(Pushing away white herons, ducks, and carp,
the nutria goes his way in the River Sasagase.)[73]

A few nutria have been reported in South Korea, but their date of introduction and current numbers are unknown.[74] Maryland officials who were active in the Delta National Wildlife Refuge (DNWR) nutria-eradication program hosted delegations from South Korea and Israel to help those nations learn about nutria and how to control them, and delegates from South Korea also traveled to Louisiana for nutria-related professional development.

## Bon Appétit

Most nutria harvested during the CNCP are not used for fur or meat. For example, only 26 percent of the 308,160 nutria trapped during the 2007 season were sold. The rest were buried, thrown into heavily vegetated areas, or sunk into a watery grave.[75] A September 2009 article posed an all-important question with its title—"Can Nutria Fur Make a Fashionable Comeback?"—and cited the following statistic: "Of the 334,038 nutria harvested during the latest season [2008–9], . . . only 11 percent of hunters used them for meat or fur." As for the rest, the article stated that "[89 percent] were left to rot in the marsh."[76]

With fur trading disfavored by the public, an idea to eat nutria before they ate the wetlands was perpetuated by the LDWF in the late 1990s before providing the bounty. A formal initiative advertising human consumption of nutria was also well supported. Armed with a $2.1 million federal grant to eradicate nutria by increasing their commercial potential, Louisiana's Nutria Harvest and Demonstration Project (NHDP) introduced eating nutria as one component of a multifaceted response to increasingly problematic nutria populations.[77] But despite well-planned efforts that included brochures and online materials with recipes that called for nutria meat, the campaign did not fetch nearly as much public support as the bounty program. While the NHDP elicited some curiosity toward nutria consumption, it seemed to fail with regard to making nutria an everyday meat along the lines of pork, beef, or turkey. Some folks say that result came because funding was skewed toward bounty provisions; others think that nutria consumption will never catch on because of the social stigma associated with eating a rodent. Either way, the nutria meal promotion ended in 2003.[78]

The NHDP was based in historical precedent, as some folks in rural Louisiana have eaten nutria for decades. Additional precedent for human consumption of nutria meat comes from the former Soviet republics of Kyrgyzstan and Uzbekistan, where nutria (called *нутрия* in Russian) are farmed on private plots and sold in local markets as a poor man's meat. According to one story from the country formerly known as East Germany, nutria farms were not unusual until the fall of Communism and German unification. The new government did not subsidize or sponsor the farms, and a whole generation of working-class East Germans who grew up eating nutria because it was an inexpensive and widely available source of protein lost out on eating nutria when the farms went out of business.[79]

Nutria meat does not fit the "tastes like chicken" cliché, but the taste, texture, and appearance of nutria meat is similar to dark turkey or rabbit. Very few pathogens are known to contaminate nutria meat, so it is a generally safe food. Nutria meat can also provide excellent nutrition. It is leaner than turkey (1.5 grams of fat per 100 grams of meat versus 2.6 grams) and beef (1.5 g/100 g vs. 26.6 g), has more protein than those meats (22.1 g/100 g vs. 21.8 g/100 g for turkey and 16.6 g/100 g for beef), and is lower than most meats in cholesterol.[80] But the nickname *swamp rat* reminds folks that nutria are a distant evolutionary cousin of a species that is closely associated with roadkill and disease. Thus, attempts to establish markets for human consumption of nutria meat have been largely unsuccessful. This trend is particularly salient in North America, where rodents are generally not regarded as food because of a cultural "ick" factor. In other words, putting a rat on a dinner plate has not caught on.

The notion of using nutria as animal feed has also gained limited momentum, but a few minor successes are worthy of mention. Nutria meat is sometimes sold to mink ranchers or used as a feed supplement for the alligator farming industry. At some point, harvested nutria were used to feed exhibit animals at the Audubon Zoo.[81] Because it is a nutritious and high-protein food, some people use nutria meat to feed animals that they own, such as falcons or dogs. A doggie biscuit made from nutria harvested by licensed CNCP hunters in the Barataria-Terrebonne Estuary and sold under the brand name Marsh Dog is gaining popularity. As with McCree's clothing company, BTNEP grant money helped Marsh Dog's company start up, and the lifelong-Louisianan owners were recognized with a Business Conservationist award at a Governor's Conservation Achievement event.

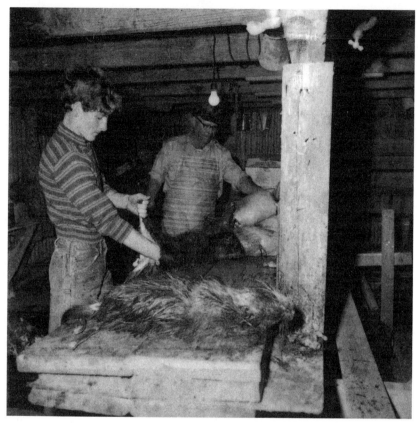

Festival. Men skinning an animal, probably a muskrat or nutria, at the 1970 Fur and Wildlife Festival in Cameron, Louisiana. Courtesy of State Library of Louisiana.

Butchered nutria is not available from a supplier, nor is it processed commercially at any USDA-approved facilities. With no formal processing and storage infrastructure to prevent spoilage or contamination of nutria meat, and no sign of a burgeoning market that would make such an operation worthwhile, logistical deficiencies continue to prevent the commercial sale of nutria meat. Previous incidents of contamination from toxicants used to control nutria populations and lead poisoning of farm-raised alligators from bullet fragments processed with nutria carcasses have probably not helped market for nutria meat either.[82] Thus, folks who want to eat nutria or feed nutria to their animals must hunt and butcher nutria themselves, or find the meat through informal channels. For most people, this is an additional deterrent to eating nutria meat.[83]

Given the stigma in North America attached to eating "swamp rat" and the general public opinion that nutria meat is not fit for human consumption, suggestions about feeding nutria to the homeless or incarcerated have also gained little traction.[84] Most progress toward using nutria meat has occurred on a small scale in rural areas of Louisiana where folks are culturally more receptive to novel foods. For example, the LDWF drummed up interest in nutria meat with a Nutria Fest at the Louisiana Nature and Science Center (LNSC) in New Orleans, which attracted a few hundred people and allowed cooks to serve nutria-based dishes. The annual Louisiana Fur and Wildlife Festival (LFWF), known as the "coldest" festival in Louisiana because it takes place during winter, honors the state's resource-utilization tradition with a muskrat- and nutria-skinning event, exhibiting an important skill for any aspiring nutria cook. Louisiana Seafood Exchange in Baton Rouge used to process and sell retail nutria meat; Bocage Market, also in Louisiana's capital, sold nutria meat for a brief period. Bear Corners Restaurant in Jackson, Louisiana, used to serve nutria dishes for lunch. Curious diners have also tried nutria meat at the Louisiana Food Service Exposition (LFSE), where the LDWF sponsored a nutria booth and cooking demonstration.[85]

National attention for eating nutria resulted from a visit to the Third Coast from television star Andrew Zimmern, who enthusiastically consumed nutria stew on-air during a 2013 episode of *Bizarre Foods*. While Zimmern's antics showed that nutria were acceptable for human consumption, the show occurred long after the LDWF promotion ended a decade before. Thus, eating nutria within the forum of a television show called *Bizarre Foods* might have perpetuated the attitude that eating nutria is too offbeat for most of the American population. Internationally, the nutria-meat movement garnered some interest when nutria was promoted at a Tokyo food exposition.[86]

Because human nutria consumption has failed to catch on in the United States, using nutria for meat is not widely considered to be a comprehensive solution for the nutria invasion. But alligator meat has been on Louisiana restaurant menus for decades even though diners were originally resistant. Perhaps alligator meat is more popular than eating nutria because of the "top-of-the-food-chain" feeling garnered from eating an aggressive, apex predator. Even so, the American acceptance of alligator meat means that there still might be time for nutria consumption to become part of a multi-angle nutria control program. Or maybe future businesspeople will finally develop a market for nutria meat outside of North America, where folks are less squeamish about what goes on their plate. Meanwhile, some bayou dwellers believe that the best opportunity for nutria meat lies in enlisting Louisiana's finest chefs to help promote nutria as not only food, but also as a cuisine. "Try it," they tell me. "You'll like it."[87]

✦ ✦ ✦

From the other side of a dimly lit wooden table, tucked into the corner of a coffee shop on the outskirts of Baton Rouge, Chef Philippe Parola flips deliberately through a colorful presentation binder. He adjusts the collar on a golden polo shirt that complements a nicely pressed pair of slacks as he enthusiastically tells me about his life's work—cleansing the environment of invasive species by bringing the harmful organisms to our dinner plates.

"If you can't beat 'em, eat 'em!" Parola states emphatically with his shoulders back and while pointing confidently with his index finger. "We have an opportunity here to nourish our nation, bring an excellent source of protein into our food supply, and get rid of an invasive species at the same time. It's a win-win."[88]

Chef Philippe's work with invasive species reached a turning point in August 2009, when he was hunting for a unique fish to cook while being featured on the Food Network's *Extreme Cuisine*, hosted by television wildlife guru Jeff Corwin. Chef was on a boat looking for a native Louisiana fish called alligator gar (*Atractosteus spatula*) with a Cajun fisherman named Billy Frioux. When a barrage of giant silvery fish suddenly shot out of the water and two flipped on board, Chef Philippe had a eureka moment.

"Let's cook 'em!" he thought.

Once Chef Philippe decided to cook the fish, his training as a chef and outdoorsman kicked in.

"Being a fisherman I knew that all jumping fish are bloody and need to be bled," recalls Chef Philippe nonchalantly on his website (www.cantbeatemeatem.com), "and being a chef, I knew to cut their tail fins and place them on ice. After cutting and skinning the [fish that landed in the boat], to my surprise the meat was white as snow. Excited with this discovery, I fried a few strips and learned that this fish was excellent eating."

The two fish that jumped into Chef Philippe's boat were Asian carp (family Cyprinidae), a nonnative species that has spread rampantly throughout the Mississippi River basin and is threatening multi-billion-dollar fishing industries from Michigan to Louisiana. For the past few years, Chef Philippe and his business partners have been working on what he feels is the only sustainable solution to the Asian carp crisis—commercial harvest for human consumption in domestic markets. Parola has launched a campaign to promote Asian carp under the trade name Silverfin, and he hopes to build an ecofriendly food-processing plant that specializes in producing Silverfin products.

For Chef Philippe, cooking up invasive species is not just a job—it's a lifestyle, and a passion. As I sip a black, dark-roast coffee and listen to Chef lay out his Silverfin plans, it is obvious that he is passionate about food, the environment, and his family. Parola's positive attitude and charisma makes him all the more convincing as he gestures confidently and expresses his thoughts with concision and candor. He tells me about being destined for his current occupation long before the Asian carp serendipitously flipped into his friend's boat. Growing up in France before he came to the United States in the 1980s, Chef Philippe forged a deep connection with nature as a young orphaned child. Like other kids, Philippe Parola went to school and dutifully completed homework or daily chores. But Philippe also hunted, fished, and foraged for mushrooms and berries to put food on his table. He therefore grew to understand nature's central role in sustaining life, a value that he has passed on to both of his daughters.

"My daughters enjoy fishing and being in nature," the coffee-less Parola explains, "and I want to make sure that they will enjoy their relationship with nature for the rest of their lives. I want to ensure that invasive species do not destroy our natural world."

Parola's appreciation of nature's bounty inspired him to become a chef, and he is now an internationally recognized authority on culinary arts and food marketing. He has prepared cuisine for a wide spectrum of people, including royalty, US presidents, fishermen, and the homeless. Top international media outlets like the *New York Times, Taiwan News, Paris Express,* Food Network, CNN, Fox News, CBS, *ABC World News,* ESPN, National Geographic, PBS, and BBC have featured his work.

Chef Philippe's attempt to rid the Mississippi River basin of Asian carp is well-known, and so are his efforts to feed nutria to the public. In the 1990s, when Parola was executive chef and owner of Bear Corners Restaurant and the Louisiana Culinary Institute (LCI) in Jackson, Louisiana, he teamed up with the LDWF to launch the challenging and somewhat unpopular campaign to market nutria for human consumption. "The only thing that's wrong with this little boy [the nutria] is that he's got a tail," was one of Parola's mantras.[89] For a few years, the media paid attention to the campaign, but the excitement eventually died down. Once I realized Parola's connection to the LDWF program, I wrote to him so that I could hear everything possible about human consumption as a potential solution to the nutria invasion:

I'm currently writing a book on nutria for the University Press of Mississippi and during my research I have come across your efforts to combat the invasive populations of nutria by creating interesting dishes with their meat for human consumption. I was hoping to talk to you about your work with nutria, as I think readers will be very interested to hear your ideas. I am also tentatively planning to come to Louisiana [this fall] and I was wondering if there was a way to taste nutria dishes at one of your restaurants or somewhere else while I am there. Would you have some time for an interview next week? Hope to hear from you, and thanks.

Best,

Theo Manno, PhD


True to the form he exhibited at the coffee shop, Chef's e-mailed response came from his cell phone with promptness and enthusiasm. During subsequent contacts, Chef said that he would contact some people, including "top scientists," to get them involved with acquiring a nutria for me to eat.

*Wow,* I thought.

A few weeks later, after some more e-mail and phone exchanges with Parola, it became apparent that obtaining a nutria for a Friday night dinner in Baton Rouge might be feasible, although the fee to cover Chef's costs was a bit of a stretch for a biology instructor. Willing to pay the higher-than-usual cost of the unusual meal, and remaining enthusiastic about the prospect of eating swamp rat, I kept at it:

Hi Chef,

Just wanted to check in and see if you were able to come up with a nutria for me to eat (and write about) for my book with the University Press of Mississippi. If so, we could get together [next week].

If not, I understand, and I thank you for your help. Perhaps I could just interview you instead, and if nothing else, I can write about how hard it is to get a nutria that you don't hunt yourself. I might like to put two or three of your nutria recipes . . . in the book too . . .

So let me know what you think, and thanks again for looking into this.

Theo Manno

P.S. Is there anywhere you know of where I can purchase nutria meat?


Of course I knew the answer to the last question was no, but Chef's people were still working on acquiring a nutria, which he informed me would be served like a turkey on Thanksgiving, with the whole nutria on the serving plate.

My mouth started to water each time I thought of the dinner. But three days later, as I flew into New Orleans, I realized something.

I was a little bit scared.

*Eating swamp rat*—really?

I had already eaten alligator, rattlesnake, rabbit, squirrel, venison, bison, elk, and probably some other meats that are usually not available at a local grocery store. *This time*, I thought, *maybe I'm going too far*. I could not put my finger on the reason, and my concern was probably irrational. But the bottom line was that I was worried.

*I'm from New Jersey. I'm not supposed to eat stuff like this.*

So it almost came as a relief when, as I was watching the Saints-Falcons football game in my New Orleans hotel room and announcers were reminiscing about Gleason's blocked punt from seven years ago, Chef called to tell me that he did not yet have a nutria for the dinner that we planned. "The hunters," he said, "haven't been able to shoot one. [Nutria are] not in season … the grass is quite tall this time of year, and it's very tough to get one."

I paused for a moment, and my eyes grew wide. Chef was sincere with me throughout the process, but it never occurred to me that hunters would try to shoot a nutria upon my request. I thought the nutria I would be eating was a casualty of the bounty program and already killed, maybe sitting in someone's freezer. And then I remembered the lone nutria at the Audubon Zoo earlier that morning, the one who seemed to grunt at me as I left. I thought of all the endearing behaviors nutria have, and how their ancestors never asked to be brought to our continent. And then I shifted from being worried to becoming upset. Here, a fascinating creature deep in Louisiana's wetlands that was acting only as nature intended was about to be shot to death—at my behest.

To be clear, I could never kill a nutria myself, for dinner or a bounty. Even though they are hurting the environment, it is not in me to shoot an animal. But I am also not vegetarian. For a while I tried to reconcile this, wondering ad nauseam about whether I should go vegetarian (or vegan), and halfheartedly trying it on a few occasions. But after the diet made me feel sluggish, I realized that I could not keep up a vegetarian lifestyle. So why was I troubled by the prospect of eating swamp rat? Probably because I am a suburbanite, and there has always been a certain abstraction to my meat eating. After someone else far away slaughters the animal, I go to a grocery store and purchase the meat in some sort of intangible form, like burgers or ground meat. This means that I only vaguely realize what I have purchased—a dead animal—and I hardly think about it. The animal is also usually considered to be livestock that is raised specifically for meat, rather than a species widely regarded as wildlife. Even in cases where the animal is served whole, such as a Thanksgiving turkey, I have eaten the animal often enough that I am anesthetized to it once being alive.

Perhaps I should reconsider the idea of any novelty meat tasting. I already eat meat, and this is hard to change now after growing up that way, but if I really think about eating a dead animal per se, the notion is repulsive, and expanding into new types of meat beyond what I already eat feels distasteful. But this line in the sand seems irrational—I know that I am just as responsible for an animal's death whether a hunter fills my order or I bought the butchered meat at a grocery store. So it appears that a person is brought up in a certain way, and that was how I felt—right, wrong, or indifferent. I suspect that the same thought process applies to folks who hunt and trap in today's anti-fur climate.

Either way, it seems that Parola has a different philosophy than mine.

"Don't feel bad," Chef told me, "we'll see if we get a nutria tomorrow. But if not, we'll still meet in Baton Rouge, okay?"

"Okay."

A long pause followed before I spoke again.

"Uh . . . is eating this nutria . . . this is legal, right?"

[*Laughter from Chef*] "Huh? Yes . . . of course!"

"All right . . . I mean, I don't want to do anything illegal."

"Well, neither do I."

Not knowing what else to say, I thanked Parola and hung up, unsure as to whether I was rooting for or against the hunters who were heading into the swamps the following morning to dispatch a nutria. Meanwhile, the Saints scored another touchdowns to take the lead against the Falcons.

*What am I upset for? They're destroying the state!*

I awoke the next morning and downed the second half of the seafood special that I ordered at a Cajun-themed roadside establishment the night before. As I washed the fried fish down with pecan coffee, I contemplated the possibility of consuming swamp rat for dinner after already eating an unorthodox breakfast. And while I think I would have overcome my apprehensions and eaten the nutria, we will never really know because the hunters failed again that morning.

Thus, the lunch meeting at a strip mall on the eastern fringes of Baton Rouge was planned instead. "Tell me more," I ask Chef Philippe, "about how you got into this line of work." Upon hearing this prompt, Chef immediately launches into the compelling story of how he was largely responsible for making alligator an acceptable food source in the Gulf Coast region.

"I started marketing exotic game in 1985 with my friend Egon Klein, who moved to Louisiana in the 1970s to buy alligator skins from trappers. He shipped them to Italy for tanning and processing."[90]

"Egon's business started to decline in the late 1970s because of the animal rights movement that started in Europe. No one wanted to wear furs or skins from wild animals. One day during lunch hour, Egon came to my restaurant with a little ice chest full of alligator meat. He asked me: 'Chef, can you create alligator meat recipes so that I can sell the whole alligator instead of just the skin?' I told him to come back the next day and I would have a few dishes for him to taste."

Together, Chef Philippe and Klein launched an alligator meat campaign that featured dishes like precooked smoked alligator loin and marinated tail meat for alligator beignets. They brought their idea to French epicures at the Salon International de l'Agroalimentaire (SIAL) in Paris, the largest food innovation observatory in the world. For the first time, alligator meat from Louisiana was on the international market as a premier exotic delicacy, and Parola's recipes caught on from there. Today, alligator meat is sold in many Louisiana eateries. "It took 10 years of policy change to have the meat cross state lines, but now the price runs around $18 a pound," Parola tells me proudly.

Chef's attempts to sell nutria meat, which did not progress as well as the alligator campaign, followed an unexpected shift in career focus. When the United States entered the Iraqi war, the French president refused to offer alliance. French chefs all over North America suffered the backlash. As more and more pro-war folks in Louisiana boycotted Chef Philippe's restaurant, he was forced to sell it.

"But I never considered quitting the food business," says the man who brought alligator to our tables. "It was an opportunity to focus on other goals, like the nutria campaign."

Along with three other top chefs, Chef Philippe tried to convince consumers that nutria meat is high in protein, low in fat, and healthy to eat. Chef says that while he was executive chef and owner of Bear Corners Restaurant and the LCI in Jackson, he served *culotte de ragondin à la moutarde*—nutria with honey-mustard sauce—as part of the regular menu. Anecdotally, Parola insists that about one-fifth of diners at the restaurant chose the dish without being dared to.

With the support of LDWF biologists Noel Kinler and Edmond Mouton, the chefs cooked nutria stews, nutria soups, roasted nutria, and grilled nutria to serve at social functions. Collaborating with Cree McCree, Chef

Philippe cooked for an event at Bizou Restaurant on Saint Charles Street in New Orleans that featured a nutria fashion show while over 300 folks dined on nutria. According to Chef Philippe, nutria meat was enjoyed by the vast majority of folks at the event.

"But there were some issues with marketing nutria meat long-term," Parola tells me.

Like what?

"First, there was the usual issue with some consumers regarding the psychological outlook that nutria resemble oversized rats. Then, we could not get USDA approval to sell the meat for human consumption because herbivores have to be killed in a slaughterhouse under FDA supervision. USDA also has guidelines for mammals that are very strict and we cannot ship them across state lines. The meat has to be brought live to a slaughterhouse for inspection, and there's a serious concern about bacterial contamination."

"So why wasn't that the case for alligators?" I ask, intrigued.

"They're classified as seafood," Chef says. "With herbivores, the regulations are different. Even when I went over to China and Japan, they wouldn't take nutria without an FDA regulation stamp.

"But the regulations don't always make a lot of sense. For instance, anyone can go out here," Parola continues, pointing in a general direction south of the highway I exited to access the coffee shop, "and shoot a wild boar, and serve it to a hundred people at a tailgate party before the LSU [Louisiana State University] football game. But as a professional chef, with my knowledge and training . . . I know my situation is commercial and that's not, but how is what I would cook up worse than that?"

When the late Jefferson Parish sheriff Harry Lee decided to use nutria as practice targets for his officers, Parola's cause took another hit. Shortly thereafter, local media reported that nutria were seen in New Orleans gutters, rendering them a nuisance species that was hard to see as edible wild game. According to Parola, all efforts to sustain a nutria market collapsed within days of the headlines.

"So how can we make it work?" I ask. "What do you need to get people eating more nutria?"

"The first thing we have to do is create an educational engine to talk to kids and lawmakers about the opportunity we have were with the nutria. We have hunger in the USA, and we have to face reality—one in five children face hunger. Nutria are a source of natural protein that could help a school system or soup kitchen . . . plus, so many people here in Louisiana love the outdoors and want to fish or hunt and leave an impact on the next generation. We have to convince folks that we can get rid of the problem, and bring food to table. And once policymakers and regulators realize that we want to work with them to control the nutria population and feed the hungry, we have to change the guidelines so that we're not wasting the food source.

"This bothers me," says the Chef, shaking his head from side to side instead of up and down, "that you can make nutria jerky, nutria gumbo, meat pie." As Parola continues, I think of Bubba describing all of the uses for shrimp in *Forrest Gump.* "We want to fight obesity, and the nutria are resources that can be easily processed into food sources that can be served to kids in lunches. But we can't get it together. I mean, thousands of people can be fed on this."

Chef's point is clear and he makes a compelling argument about using the whole nutria. But I wonder how many folks will listen. I do not have a picky palate, but I still fought my own hesitations about eating nutria, even if prepared by one of the best chefs in the world. So I start looking around the coffee shop to demonstrate a hypothetical.

"You think that if I turn around here," I say, "and start asking people if they want to eat swamp rat, that they'll jump at the opportunity?"

"No, probably not," Chef answers, "but if I walk around the room and say, 'I have nutria meat—will you try some?' Then I think most people here will. There's a way that you market the product. But you have to take the right product and advertise it correctly. For instance—convenience . . . that's one factor now. So, maybe nutria jerky. Or you can market nutria as exotic, or a novelty, or delicacy."

It is easy to tell other people to eat nutria. "So," I ask Parola, "would you eat nutria yourself?"

"Of course, we have to let people know that I eat it and the best chefs in the world eat it," he says, "because if we didn't eat it ourselves, then why should other people try it? We've got to be clear that we're not lying about it. Let me tell you, if nutria was not good, then I wouldn't serve it," he says confidently.

"Now, it took six years with the alligator, but we were totally successful. We've got the blueprint from the alligator campaign. So we can show lawmakers a plan for what works, and take that path to success and use it for the nutria campaign. But you have to do your homework on something like this, and homework costs."

The seemingly simple task of deciding on a name for a food product, Chef tells me, can be a long and excruciating process that takes years. When Chef came up with the name Silverfin for his Asian carp products, teams of copyright lawyers researched the records of international regulatory agencies to make sure there were no trademark infringements. Parola also

# Nutria Recipes by Chef Philippe Parola

### Ragondin à l'Orange
Ingredients

*Mirepoix*
1/3 cup chopped celery
1/3 cup chopped carrots
1/3 cup chopped onion

*Nutria*
2 hind saddle portions of nutria meat
1/2 cup brown sugar
1 cup orange juice
2 tablespoons vegetable oil

*Bouquet garni*
1 bunch fresh thyme
1 bunch parsley
bay leaves
2 cups white wine
1 tablespoon soy sauce
Orange zest, minced
Salt and pepper to taste

Place oil, mirepoix, and bouquet garni in a pan; set aside. Rub each hind saddle with brown sugar and salt and pepper to taste. Place hind saddles on top of other ingredients in pan. Place, uncovered, in a 350-degree oven for 15 minutes.

Remove from oven and deglaze with white wine, soy sauce, and orange juice. Cover pan with plastic wrap, the cover again with aluminum foil. Place back into oven for 45 minutes to one hour until meat is tender. Break meat off bones. Place on plate, then garnish with vegetables, sauce from pan drippings, and orange zest. Makes 4 servings.

### Culotte de Ragondin à la Moutarde
Ingredients

*Mirepoix*
1/3 cup chopped celery
1/3 cup chopped carrots
1/3 cup chopped onion

*Bouquet garni*
Fresh thyme
Parsley
Bay leaves

*Sauce*
1 cup demiglace
1/2 cup heavy cream
2 teaspoons Dijon mustard

*Nutria*
1-1/2 tablespoons vegetable oil
2 hind saddle portions of nutria
4 tablespoons Dijon mustard
1 cup white wine
1-1/2 tablespoons olive oil
Salt and pepper to taste
1/2 teaspoon crushed rosemary
2 cups water

Into saucepan, add demiglace, cream, and mustard, stir well, and reduce on medium. Heat for 5 minutes. Season to taste.

Place oil, mirepoix and bouquet garni in a pan; set aside. Rub each hind saddle with 2 tablespoons Dijon mustard and rosemary. Place hind saddles on top of other ingredients in pan. Place, uncovered, in a 350-degree oven for 15 minutes. Remove from oven and deglaze with white wine. Add water to pan. Cover pan with plastic wrap, and then cover again with aluminum foil. Place back into oven for 45 minutes to 1 hour (until meat is tender). Break meat off bones. Place on plate and garnish with vegetables, sauce, and/or pan drippings. Makes 4 servings.

Source: Can't Beat 'Em, Eat 'Em, www.cantbeatemeatem.com; used with permission from P. Parola, personal communication, October 16, 2015.

needed advisory boards, scientists, nutritionists, and doctors to advise him and provide support for the campaign.

For marketing meat like nutria where people have ingrained misgivings about its consumption, the product name is crucial. Parola was using the French name *ragondin*—pronounced *RAH-goh-DANH*—to market nutria as a delicacy. But some people think that *ragondin* is too foreign sounding and that sales of nutria will rise if marketed to a more general audience with a down-home name like *bayou rabbit*.[91] Parola seems willing to try several marketing approaches to see what sticks.

"No book or curriculum exists on cooking and marketing nutria—it's trial and error. There's a learning curve, a very expensive learning curve," says Parola, "and that's with my knowing a lot about the industry. But from the public's standpoint, it's more expensive to leave nutria in the wetlands than fail to take advantage of a food source that's costing us in damage."

Parola is positive about human consumption of nutria, but not everyone shares his view. One example of a *ragondin* dissident is food editor and cookbook author John DeMers, who finds nutria's flavor largely nonexistent and its texture "mushy and vegetable-like."[92] DeMers does not think that nutria will be the next alligator, and considers the only potential market for nutria meat to be overseas, where people are not particularly fastidious about what they eat. But even DeMers admits that the dishes he sampled at the Bizou event were very good, because the meat's flavor (or lack thereof) was masked by the many wonderful textures and flavors created by Parola and his team of chefs. His favorite was a dish with nutria meat wrapped in crunchy lettuce leaves with soy-ginger sauce.[93]

Naysayers notwithstanding, Chef Philippe still believes that nutria can appear routinely on American dinner plates, and that if it can happen anywhere, it can happen in Louisiana. Historically, folks in Louisiana with the simple need to put food on the table took whatever potentially edible organism was available and threw it into the pot. One of the results was gumbo, a heavily seasoned soup or stew that combines meat or seafood with a sauce or gravy. Another result was the ubiquity of crawfish-related dishes—crawfish used to be considered inedible. Parola cites the long-standing tradition of hungry and resourceful Louisianans searching for food and the wealth of culinary talent along the Gulf Coast as potential rationales for a nutria meat campaign. "We put crawfish on the table, we put alligator meat on the table, we put turtle meat on the table. Consumers are looking for new stuff, something out of the ordinary. And nutria meat is out of the ordinary."

Indeed, Louisiana's fascination with new and exotic meats during the 1980s and '90s seems to set a precedent for Parola's idea to make nutria a commercially available food source. For example, another top New Orleans chef with a Cajun-inspired repertoire named Paul Prudhomme stopped redfish (*Sebastes* spp.) from overpopulating Louisiana when he started preparing "blackened redfish" with a dense pepper char during the 1980s. Prudhomme was also involved in the campaign to market nutria meat as a delicacy, saying that nutria was particularly marketable as an appetizer and very tasty when deep-fried and served as "popcorn nutria."[94] During the course of my research for this book, I contacted Prudhomme for an interview but received no response to my e-mail. During my meeting with Chef Philippe, I learned why. Sadly, Chef Prudhomme passed away just a week or two before I came to Baton Rouge.

"A journalist once said that the redfish outsmarted everyone except Chef Prudhomme," says Parola. "What a tremendous loss to the kitchen. The man almost drove the redfish to extinction with a single recipe."

"Don't underestimate the power of cooking."

Even though thousands of nutria have been wiped out by recent hurricanes, their presence in the wild is still causing environmental harm. Thus, wildlife officials can only hope that Prudhomme's success with cooking redfish can be replicated to trounce nutria from Louisiana's wetlands. Meanwhile, many locals continue to envision nutria as roadkill, vermin, or maybe a sports mascot. Creating a year-round supply of nutria meat remains a profound issue, as demonstrated by the inability to fill my order outside of the winter hunting season. Even Chef Prudhomme, who was an ardent supporter of the nutria meat program, dropped out of the nutria-marketing effort several years ago after he experienced difficulty getting fresh meat outside of hunting season.[95] And almost no one wants to address the supply issue by revisiting farm-raising of nutria, which is largely to blame for our current predicament. But in spite of these challenges, Parola is a true believer in bringing nutria to the plate as a long-term, if not comprehensive, solution.

"The bounty was decided on as an immediate solution to the problem," he says, "but cooking the nutria is a more long-term, sustainable solution. Maybe it's not on the federal docket right now because hurricanes have killed thousands of nutria. But if that [bounty] incentive gets cut off and federal funding is gone, we'll go back to square one again. And you and me, we're paying for it. Look at what's going on with the Asian carp [in the Great Lakes region]. We're going to build a barrier for $35,000 a day to keep them out of the lakes instead of spending that money to fish and eat them. We've got to put a value on these resources, and get hunters and trappers to come out and make a living on it. But when you go for the quick Band-Aid, and then you don't pay the fishermen enough money to fish, then you're slap-ping the hand that feeds you.

"If we use the nutria for food, it can get rid of an invasive species while creating thousands of jobs and boosting the economy. We'll bring the prod-uct right to the consumer."

Unprovoked by my questions, Parola continues excitedly as I type calmly.

"People ask, 'Well, how much am I going to be paid?' If you see some-thing like this through, if it's funded well, if the idea is presented right, then you will be paid. And not just with money. Because if we don't take care of this [the nutria problem], then future generations will have to pay. Instead of eating them, the bodies are left to rot. It's just another thing to show us that we're in the most wasteful era of humankind. We're living in an age with so much greed, and we're smarter than this, I know we are. We're not going to achieve things without struggle, without giving. Imagine, if everyone can make a difference with what they have instead of worrying about them-selves, if we stop our greedy type of thinking, then how can we take care of our resources and our children?"

On that note, Chef and I leave the coffee shop, say good-bye, and shake hands as he asks me to be open-minded as I conduct my research. We've been talking for hours and my light lunch was finished early into the inter-view. Hungry for an early dinner, I walk across the street to an upscale gro-cery store, bordered by the swamplands on the cusp of eastern Baton Rouge.

I am surrounded by the unique cuisine of Louisiana, and looking for food that I cannot obtain anywhere else. While walking down the aisles, I see seafood salad, the popcorn crayfish that was once considered inedible, and the alligator gumbo that Chef Philippe pioneered over 30 years ago. *Will we ever see popcorn nutria on these shelves?* I thought as my eyes went to a thick pie under a long, clear piece of glass. *What if nutria meat actually catches on one day, like the alligator meat?*

*Will we be back to farming nutria?*

With no answer to these questions, I refocus on the baked goods. "Is that pecan pie?" I ask the young lady behind the counter.

"Yes, it's bourbon pecan pie," she answers.

"*Bourbon* pecan pie?" I respond, without using my internal "edit" button first. "Is the bourbon really necessary?"

The addition of alcohol to the pie seems like a bit much for my stomach as I settle for a slice of the nonalcoholic version that I notice three pies to the left. As I look for a main course, the alligator gumbo seems appropriate. I fill a small container with the creamy reptile soup and pay for it along with the piece of pie at the register. Inevitably, the cashier asks me if I would like anything else.

"Do you have any Tabasco?"

They do, so I seize the red elixir in the small bottle with the diamond-shaped logo, screw off the top, and combine McIlhenny's sauce with Parola's alligator meat. I walk outside to the patio and select a table for my exemplary Louisiana meal, looking out in the direction of the swamps, thinking about Chef Philippe's words.

*What will this all look like in 10 years? What about 50 years?*

*What will be our statement?*

The sun sets on Baton Rouge with a pinkish hue over the horizon, then darkness. From the direction of the swamps I hear frogs croaking and crickets chirping; behind them in my thoughts are the words of scientists, historians, policy makers, chefs, clothing designers, trappers, and other concerned citizens. Then, in the distance, a sound emerges that is louder than all of their arguments put together—a piglike grunt, probably from a nutria, travels loudly across the bayou. Somewhere in the wetlands it is probably traipsing through vegetation or tending to its kits, living in its home away from home.

# Chapter 7

# Afterword

Resource managers don't manage animals or land. They manage people.
—Author unknown[1]

In August 2008, I completed a doctorate from Auburn University and drove west to accept a teaching position in Arizona. When New Orleans appeared after five hours on the interstate, I stopped and stayed overnight to check on how the city was faring after Hurricane Katrina. It had been three years since my last visit to New Orleans, just before Katrina's wrath flooded her streets.

Although I saw a resilient city, I also observed a community that was struggling to recover its infrastructure. Streets that were once hectic seemed empty and were speckled with abandoned houses. The notorious overpass and repaired Superdome, where so many folks waited out the storm in deplorable conditions, stood eerily as the *Rebirth* statue glimmered in the summer haze. But the most haunting sights in New Orleans were the "X-codes"—officially called "search codes" by the Federal Emergency Management Agency (FEMA) and sometimes called "Katrina crosses" in the colloquial—painted on buildings and houses, standing stoically as narratives of the hurricane and reciting a horrific tale of destruction. I read later that four-fifths of the city's structures were crudely spray-painted with an X-code, which featured a scorecard of letters and numbers in each makeshift "quadrant" of the X. In the leftmost quadrant was shorthand that identified the rescue squad; at the top, a time and date when the team arrived. The right-hand quadrant featured a summary of hazards within the house. Then, in the bottom of the X, rescue workers listed the number of people found inside. The first number represented those found alive; the other, those who were not.

Just a few weeks before Katrina slammed into the Gulf Coast, a unintentionally prophetic study conducted at the Massachusetts Institute of Technology (MIT) and published in the high-impact scientific journal *Nature* concluded that tropical cyclones were lasting longer than before and had become 50 percent stronger over the past 30 years. The trend correlated with a rise in sea surface temperatures, which the author suggested might be increasing more quickly than atmospheric temperatures.[2] While the author was hesitant to generalize his conclusions, the study's implications were obvious. Indeed, most scientists maintain that rises in sea surface temperatures during the past 30–50 years are a signal of global warming. In 2014, the Intergovernmental Panel on Climate Change (IPCC) reported that this phenomenon was, to a degree of 95 percent certainty, mostly caused by increasing concentrations of greenhouse gases in the atmosphere from automobiles, industry, and other anthropogenic activities.[3] "That's their [the scientific community's] conclusion, not mine," said the *Nature* study's author during an interview with *National Geographic News*, "[but] it would follow reasonably well from this metric that the upswing [in intensity] . . . is a result of global warming."[4]

Like hurricanes, it seems that nutria are also on the move as weather warms. Researchers with the US Geological Survey (USGS) and Portland State University (PSU) in Oregon found that during a recent string of mild winters, nutria populations in the United States expanded northward. According to the study, this trend could continue over the next 40 years if left uncontrolled.[5] The researchers focused on the current and potential distribution of nutria in Washington and Oregon, where nutria occur west of the Cascade Mountains but may spread farther inland. An increased geographic range for *M. coypus*, a species known to thrive in warm weather, could mean more damage to wetlands and agricultural fields in the Pacific Northwest, along the Gulf Coast, and even on the eastern seaboard. Preventing the hungry rodents from increasing their range has therefore become paramount among officials in these locales. "Monitoring the primary routes of potential dispersal, such as the Columbia River [along the border of Washington and Oregon], will be important for limiting expansion eastward," explained one of the study authors in an interview with *Scientific American*. Taken together with climate change models that predict milder winter temperatures, the scientists also suggested that "[by] the year 2050, we show that almost all of the [United States is] suitable for nutria [habitat]."[6]

Nutria proliferation and hurricane-related storm surges are probably more impactful because of global warming. But whether nutria are exacerbating the detrimental effects of hurricanes per se remains to be seen. While nutria feeding may lead to increased erosion when an intense storm makes landfall because the stability of grass roots that hold together soils becomes compromised, this trend is mitigated by two other factors. First, areas that are severely affected by hurricanes do not always contain high numbers of nutria. And second, storm surges often kill thousands of nutria, leaving their corpses to rot on the beach. Indeed, recent hurricanes are probably one of the reasons that nutria-related wetland loss is not as profound as it once was.[7]

What is clear is that nutria introduction should be placed among the countless anthropogenic activities that contributed to the recent economic and environmental downfall of the Gulf Coast region. It was humankind that caused saltwater to flow into freshwater wetlands by building canals. We constructed levees that compromise natural processes like deltaic cycling. We logged the swamps into oblivion. We are spewing greenhouses gases into the atmosphere that cause global warming, and probably stronger or more damaging hurricanes as a result. And as if these issues were not enough, we also introduced vegetation-denuding nutria into a land that nature never intended to be their home.

How much longer can humankind make the same mistakes before we take responsibility for our investment in protecting the environment? After all, we need the Gulf Coast wetlands for our flood control, storm protection, fisheries production, carbon storage, and water filtration. If these crucial wetland functions are compromised, then we will ultimately suffer from our thoughtlessness. For now, we have the nutria bounty program, which is probably a necessary rejuvenation of a largely defunct fur industry that mitigates the long-term effects of nutria introduction. Meanwhile, the people of America's Third Coast must look for rebirth and recovery, which means catalyzing economic activity without permanently destroying a unique environment that generates industrial opportunity. To accomplish this, folks should take some lessons from history. In retrospect, we see that the fur industry might not have self-destructed and demanded the introduction of nonnative nutria if it had progressed more sustainably. Settlers already hunted and trapped most other furbearers into near-oblivion, harvesting populations zealously without allowing them to replenish themselves. Instead of taking this hint from nature, we looked to South America for more animals to kill—and then we obliterated the large alligators that, for all we know, might have helped save us from the nutria invasion.

## Balance

Uninformed introductions of nutria gave us more than we bargained for. Even when thousands of nutria corpses are washed ashore after a hurricane, nutria reproduce so quickly that populations will almost always rebound if not controlled with hunting and trapping, and sometimes even if they are. Unfortunately, our precious wetlands have not proven to be as resilient. Thus, the story of Dixie's nutria invasion serves as a cautionary tale. Few people could have foreseen the bottom dropping out of the fur industry, nor could most folks predict that nutria would cause such widespread damage. And wildlife managers probably encouraged nutria transplanting with the idea that either nutria or muskrats would emerge as sufficient stock for the fur trade on which the state's economy traditionally depended. Nevertheless, the complexities of fluctuating markets and the tenuousness of ecological interrelationships probably should have necessitated more careful consideration by interested parties. Folks were well aware of some dangers that could result from transplanting animals between countries. But instead of proceeding cautiously, people like William F. Frakes, H. Conrad Brote, Abe Bernstein, Ned McIlhenny, and others started nutria operations even though they were ignorant of the challenges that raising nutria posed. Nutria were then spread across the southern states, usually at the behest of wildlife managers who ignored the potential impacts of proliferating nutria years down the road.

Is it easy to say, given the benefit of hindsight, that humankind could have stopped the nutria disaster if we had only read the writing on the wall? Perhaps it is. But the fact remains that, from now on, people should remember the nutria invasion and consider all political, economic, and environmental ramifications before purposely releasing a nonnative species into the contemporary landscape. With that said, should we then place blame on those who miscalculated or ignored the effects of introducing nutria? This is not an easy question to answer, because the economic atmosphere surrounding the fur industry in the 1930s through the 1960s was much different than it is now. In addition, many of the nutria introductions were well-intentioned (if not misguided) attempts to control weeds or revitalize the once economically crucial fur industry. While the nutria invasion should serve as an example of what can happen when introduced species become invasive, perhaps our energy is better spent determining how to eliminate nutria and restore our wetlands rather than placing blame for their introduction.

Culpability for the nutria invasion cannot be attributed to one or two parties anyway. Ned McIlhenny liberated nutria from his stock onto Avery

Island, but he was only the biggest celebrity to deal in nutria. Previously obscure dilettantes like William F. Frakes, H. Conrad Brote, and Abe Bernstein assumed the risk of importing or breeding nutria even though folks in wildlife circles were aware that nutria disperse easily and replicate quickly. Indeed, McIlhenny was probably only the third nutria farmer in Louisiana (after Brote and Bernstein) and nowhere near to being the first nutria farmer in the United States. Premeditated nutria introductions took place all across the county, not just in Louisiana, and these introductions often occurred under the auspice of government-endorsed programs to create stock for a fur industry that the public supported fervently. Thus, one could argue that millions of people, including anyone who participated in or supported the fur trade, played a role in the nutria invasion and the environmental damage that followed.

On the other hand, some folks might blame wetland damage from nutria on animal activists for creating a contemporary society in which wearing fur products is considered reprehensible. This connection is ironic, because almost all animal rights activists also support environmental initiatives, even though absence of the fur industry that animal rights activists despise has unquestionably led to more nutria eating the wetlands. The irony is probably not lost on anti-fur organizations, which is likely one of the reasons that the nutria bounty program has met with surprisingly little resistance.[8] Thus, like many people who released nutria into the swamps, anti-fur activists may have inadvertently played a role in the nutria invasion, but an unintentional one that most would not equate with blame per se. Failure to protest the overt killing that is characteristic of the CNCP also demonstrates recognition that a larger ethical threat results from the loss of wetland habitats for other creatures.

With everyone from fourth-generation trappers to animal rights activists exhibiting no apparent objection to providing a bounty on nutria tails, it is essentially Hands Across America against the nutria. But if hunting and trapping nutria spares wetlands from damage, then one also wonders if "ecofriendly" fur is an oxymoron or a viable concept. Another issue is how much we should hunt and trap. These are not easy questions to answer, and I remain conflicted over these issues myself.

The natural history and behavior of nutria is compelling, and I must admit that my general tendency as an animal lover is to sympathize more with the nutria that are trapped rather than the trappers who now find themselves in economic straits. On the other hand, furbearer harvests were part of a traditional way of life for Native Americans, and then settlers, for hundreds of years before nutria came to the United States. I also recognize that resource utilization continues to provide much-needed economic activity for the Gulf Coast region. Many area residents utilize coastal resources via shrimping, oystering, oil drilling, hunting, and trapping, and folks who supplement their livelihood with these activities usually implicitly value local fauna and their habitat. Indeed, studies show that hunters are more likely to favor conservation initiatives than non-hunters, probably because they require animals to harvest and an environment in which to harvest them.[9]

Many respected trappers, biologists, and wildlife managers think differently than me for valid reasons. Indeed, the organized trapping community considers nutria and some other furbearers that are often disregarded as worthless pests by the general public to be valuable resources, and they are often the first to speak out about population declines or degradation of the environment. The National Trappers Association (NTA), for instance, mentions "safe and ethical harvest" and "preservation and enhancement of [furbearer] habitats" in the mission statement on their website. We have come a long way from ruthlessly trapping species into extinction, and collaborative research done by the USFWS and trapping associations has moved us toward ensuring that trapping is done humanely, sustainably, safely, and responsibly. A recent reintroduction of river otters throughout American river systems is an example of these successful partnerships between trappers and wildlife managers.[10]

Ned McIlhenny was a prime example of this balanced, two-pronged philosophy. For example, he observed alligators at some times and hunted them at others. And McIlhenny was hardly alone in this line of thinking, as LDWF officials distributed money from harvesting licenses back into conservation programs that were providing furbearers to harvest. Officials understood the reality that conservation unto itself is a hard sell; public buy-in is more likely if there is financial value to a conservation program. Through talking to people with a different upbringing than mine, I have come to understand and accept this reality too. Unfortunately, so many species and so much wetland area has disappeared that it is unwise to risk alienating anyone who values wildlife or the environment for any reason, including the desire to hunt or trap. And when the harvested animals are destroying an ecologically important area, even if overtly shot to death or brought there irresponsibly by humans, it becomes very hard to dispute a program for their control.

Meanwhile, humankind should not create environmental problems and then scramble for quick solutions that can exacerbate the original issues. Carefully managed culling is one thing, but I maintain that situations where bounty programs and indiscriminate killing must be implemented are not ideal. Thus, informed long-term management plans to use environmental resources sustainably are in order. We should never again feel compelled to introduce exotic animals that may cause unintended environmental consequences or to rescue the environment with a bounty program.

One way to prevent the atmosphere that led to the nutria invasion is to avoid conquest and short-term profit seeking. In the interests of empire and fortune, animals like beavers, seals, bison, and muskrats were systematically obliterated, and the carnage continued well beyond the necessity for survival that existed when Native American tribes dominated the landscape. As a result, folks were compelled to hedge their bets on unknown quantities like nutria. Instead of this historically stagnant approach, future management plans should ensure that we harvest fur or meat resources in a manner that allows their populations to replenish. We should also consider more closely the potential negative effects of harvesting apex predators like alligators, because there should always be an ecological fail-safe to balance populations of fecund animals.

Are higher populations of alligators sufficient for nutria control? Perhaps not, as data on this issue remain inconclusive. So, in specific situations, maybe ecofriendly fur can exist as part of a comprehensive management plan. We must control nutria somehow to deal with the mistakes that history has already made, and harvested nutria are a resource that can be utilized. But instead, hunters and trappers are redeeming tails for the cost of a bagel and coffee. This makes me uneasy, because such activity may unintentionally promote the same wasteful philosophy that led to nutria introductions in the first place. Yes, bounty programs are effective and the best option for immediate nutria control available. But rather than wanton killing, I think that we should honor traditional resource utilization by using harvested nutria in every way possible.

Because history provides a precedent, some people figure that using nutria for fur is at least somewhat acceptable, even if they consider it immoral to wear fur from other animals. But should use of nutria corpses extend to putting them on American dinner plates? I think it does. The opportunity exists to control an invasive species and help the country's hunger problems simultaneously, and I think we should take advantage. It is imprudent to create a market in which folks are compelled to breed nutria for meat, so nutria meat should be termed a finite operation that ends when nutria are extirpated, just in case it takes off in the same manner as alligator flesh. But considering that nutria is a nutritious and protein-filled meat, I hope that folks can get over the "ick" factor regarding consumption of ratlike animals that is prevalent in western cultures and use our appetites to save the wetlands. At a minimum, I hope that we can use nutria corpses from the bounty program as animal feed in zoos or for doggie treats, and place more nutria in captive situations like the Audubon Zoo so that people can learn about the damage that invasive species can cause. Is eating nutria a comprehensive solution for their control? Probably not, as my problems with finding nutria to eat demonstrate. But as a component of an overall control program, I agree with Chef—bon appétit.

And while we are at it, maybe we should use nutria for fur as well. Make no mistake, I detest the torturing of innocent animals to make consumer products for the wealthy or chic to use for conspicuous consumption—alligators included—and no one can convince me that killing animals is a "sport." But in this unique context, where the animal is invasive and killed under a bounty program anyway, Cree McCree's mantra of "creating something beautiful" might be acceptable. Proceeds from nutria fur ventures could even be used to finance wetland restoration efforts. My only concern is that if nutria fur catches on, it could reignite international interest in all furs and lead to the massacre of other innocent furbearing animals. Unpredictable swings can occur regarding what is popular, sought after, and commercialized, and folks might not necessarily be aware that a "rat hat" worn by a passerby is made from an invasive pest. Thus, programs and companies dealing in nutria fur should make their mission clear to the public, and specify that their philosophy on wearing fur applies only to invasive animals. For now, the market for nutria fur is so miniscule that the issue of unintentionally reviving a largely defunct fur industry is moot.

When a nutria traipses through the Louisiana swamps, it walks upon a landscape changed by human hands. Whether it will see a thriving wetland ecosystem in the future remains uncertain. Meanwhile, nutria control and wetland restoration efforts play out in the national eye, as America's Third Coast continues to battle the invasive rodents in addition to the massive challenges posed by powerful hurricanes and an epic oil spill. And although these efforts have profound consequences, thousands of nutria who have somehow escaped the aggressive bounty program still roam the wetlands passively with a hazy look in their eyes, interested in little else but their next meal. They know nothing of what they have done, because what happened was our fault, not theirs.

As these indifferent nutria munch away at our coasts, my hope is that this book will elicit consideration regarding how much humankind compromised one of the most unique and important areas in the country through short-sightedness. If we are to restore America's wetlands to their former glory, we must move forward with the understanding that sometimes, less is more. Whether it is collapsing wetlands back into the sea by excessively removing oil from the ground, or spewing greenhouses gases from our automobiles, or hunting and trapping so recklessly that market demand necessitates introduction of a prolific, exotic rodent and the annihilation of its most likely predator, we—not the nutria, who have merely acted according to their nature—have ruined the wetlands for ourselves. As we move forward with coastal management decisions, the nutria invasion should remind us of history's mistakes so that they are never repeated.

# Notes

## Chapter 1. An Unnatural History

1. Wishart, *Fur Trade of the American West*, 107, 161.

2. Holm, Evers, and Sasser, "Nutria in Louisiana," 1.

3. Woodward and Quinn, *Encyclopedia of Endangered Species*, xiii, xv.

4. Holm, Evers, and Sasser, "Nutria in Louisiana," 1.

5. Lowery, *Mammals of Louisiana*, 220; N. Rich, "Most Ambitious Lawsuit Ever." The estimate of twenty million comes from O'Neil and Linscombe, "Fur Animals," but it remains unclear as to whether this estimate includes the entirety of Louisiana or coastal parishes only.

6. See, for example, Kuhn and Peloquin, "Oregon's Nutria Problem"; Link, "Living with Nutria"; Willner, Chapman, and Pursley, "Reproduction, Physiological Responses."

7. "Rat That Ate Louisiana."

8. According to a US Geological Survey National Wetlands Research Center study, coastal Louisiana has lost a wetland area the size of Delaware, equaling 1,883 square miles, during the past seventy-eight years; LaVista and Bodin, "How Are Louisiana's Wetlands Changing?"

9. Schleifstein, "Bounty Hunters Making Dent."

10. Woodward and Quinn, *Encyclopedia of Endangered Species*, xiii–xxvii, 3.

11. The National Geographic Society's preferred term is *nutria* for both the singular and plural, which is the usage I employ throughout. See http://stylemanual.natgeo.com/home/N/nutria.

12. Honeycutt, "Rodents (Rodentia)."

13. Musser and Carleton, "Family Muridae."

14. Woods et al., "*Myocastor coypus*."

15. Ehrlich, "Biology of the Nutria."

16. Willner et al., "Nutria: *Myocastor coypus*."

17. Woods et al., "*Myocastor coypus*."

18. See, for example, Ferrante, "Oxygen Conservation"; Scheuring and Bratkowski, "Hematological Values."

19. W. H. Osgood, "Mammals of Chile."

20. Ibid.

21. Carter and Leonard, "Review of the Literature."

22. Ibid.

23. Evans, "About Nutria."

24. Carter and Leonard, "Review of the Literature."

25. American Heritage Dictionaries, *Spanish Word Histories*, 79–80, 160.

26. Johannsen, *House of Beadle and Adams*, 32–33.

27. Beeton, *Brave Tales*, 510–11.

28. Brande, *Dictionary of Science*, 619.

29. De Azara, *Natural History*, 331.

30. Brochet, *La Chasse aux Canards*, 29; Costanza, *Il Tecnico Operaio*, 401.

31. Broekhuizen, *De Beverrat*, 1–9.

32. Todd, *Tinkering with Eden*, 199.

33. Quilter, *Civilization of the Incas*, 160.

34. Molina, *Saggio sulla Storia Naturale*.

35. Molina, *Geographical, Natural, and Civil History*, 1–2.

36. Kerr, *Animal Kingdom*.

37. Geoffroy Saint-Hilaire, "Mémoire"; Woods et al., "*Myocastor coypus*," 1.

38. Bounds, "Nutria."

39. Knight, *Sketches in Natural History*, 70.

40. Warkentin, "Observations."

41. Atwood, "Life History Studies."

42. Norris, "Campaign against Feral Coypus."

43. Atwood, "Life History Studies."

44. Warkentin, "Observations."

45. See, for example, Aliev, "Contribution"; Ryszkowski, "Space Organization."

46. Aliev, "Contribution"; Ryszkowski, "Space Organization."

47. See, for example, Contreras, "Bioenergetics"; Gosling, "Duration of Lactation."

48. Gosling, "Twenty-Four Hour Activity Cycle."

49. Ferrante, "Oxygen Conservation."

50. Link, "Living with Nutria."

51. Gosling, "Twenty-Four Hour Activity Cycle."

52. Link, "Living with Nutria."

53. Ibid.

54. Atwood, "Life History Studies"; Gosling, "Twenty-Four Hour Activity Cycle."

55. Warkentin, "Observations."

56. Shirley, Chabreck, and Linscombe, "Foods of Nutria."

57. Holm, Evers, and Sasser, "Nutria in Louisiana."

58. Taylor, Grace, and Marx, "Effects of Herbivory."

59. See, for example, Fuller et al., "Effects of Herbivory"; Nyman, Chabreck, and Kinler, "Some Effects of Herbivory."

60. See, for example, Willner, Dixon, Chapman, and Stauffer, "Model for Predicting"; Willner, Dixon, and Chapman, "Age Determination."

61. Gosling and Baker, "Coypu."

62. L. N. Brown, "Ecological Relationships"; Skowron-Cendrzak, "Sexual Maturation."

63. See, for example, Federspiel, "Nutria Farming"; Gosling and Baker, "Coypu"; Pietrzyk-Walknowski, "Sexual Maturation"; Willner, Dixon, and Chapman, "Age Determination."

64. Atwood, "Life History Studies."

65. Gosling, "Coypu in East Anglia."

66. Nutria density estimates vary widely. For example: 138/hectare in Oregon; 21.4/hectare in Maryland; 2.1–24/hectare in a Louisiana brackish marsh; 24/hectare in a Mississippi agriculture marsh; 43.7/hectare in Louisiana freshwater marsh; and 0.72–3.7/hectare in a riparian area in Italy; Woods et al., "*Myocastor coypus.*"

67. Evans, "About Nutria."

68. Newson, "Reproduction," *Journal of Reproduction and Fertility.*

69. Doncaster and Micol, "Annual Cycle."

70. Gosling, "Duration of Lactation"; Newson, "Reproduction," *Journal of Reproduction and Fertility.*

71. See, for example, Holmes, Illman, and Beverley, "Toxoplasmosis"; Jelinek et al., "Determination of Papillomatosis"; Howerth et al., "Survey for Selected Diseases."

72. Kinler, Linscombe, and Chabreck, "Smooth Beggartick."

73. Newson and Holmes, "Some Ectoparasites."

74. Pridham, Budd, and Karstad, "Common Diseases of Furbearing Mammals."

75. Little, "Dermatitis."

76. Considered by many furbearer biologists to be the most humane way to trap nutria, these traps are often called Conibear traps after Canadian inventor Frank Conibear, who began their manufacture in the late 1950s as the Victor-Conibear trap; Bevington, "Arctic Profiles."

77. Wade and Ramsey, *Identifying and Managing Aquatic Rodents*, 12–13, 26–36.

78. Evans et al., "Techniques."

79. Ibid.

80. See, for example, Simpson and Swank, "Trap Avoidance."

81. Fichet-Calvet, "Persistence."

82. See, for example, Coreil and Perry, "Collar for Attaching Radio Transmitter"; E. E. Hammond et al., "Surgical Implantation"; Merino, Carter, and Thibodeaux, "Testing Tail-Mounted Transmitters."

83. Nolfo-Clements, "Nutria Survivorship"; Nolfo and Hammond, "Novel Method." Even though nutria have been killed en masse within the context of bounty programs, it is generally considered unethical to increase their chances of being predated as a result of interference by scientists.

84. Haramis and White, "Beaded Collar."

85. Kendrot, "Eradication Strategies."

86. Spiller and Chabreck, "Wildlife Populations."

87. Doncaster and Micol, "Annual Cycle."

88. Callahan et al., "Microsatellite DNA Markers."

89. Schleifstein, "Bounty Hunters," 1–3.

90. Bounds, "Marsh Restoration"; Kuhn and Peloquin, "Oregon's Nutria Problem"; Hazardous Substances and New Organisms Act of 2003.

91. N. Rich, "Most Ambitious Lawsuit Ever," 1–3.

92. The funds were never appropriated for this bounty; Mouton, Linscombe, and Hartley, *Survey of Nutria Herbivory Damage*, 5–25.

93. Endangered Species Act of 1973 (ESA).

94. Holm, Evers, and Sasser, "Nutria in Louisiana," 1.

95. Keddy, *Wetland Ecology*, 2–14, 227.

96. Ibid., 5–28.

97. Ibid., 18–20.

98. Ibid., 5–8.

99. Gosselink, Coleman, and Stewart, "Coastal Louisiana."

100. Keddy, *Wetland Ecology*, 155, 264.

101. Trapping is a specific license. Over 20,000 licenses were once sold annually in Louisiana, but numbers are now down to a few thousand; Alexander-Bloch, "Louisiana Commercial Fishery."

102. Keddy, *Wetland Ecology*, 155; 264.

103. Gosselink, Coleman, and Stewart, "Coastal Louisiana."

104. Ibid.

105. N. Rich, "Most Ambitious Lawsuit Ever."

106. Keddy, *Wetland Ecology*, 155, 264.

107. Woodward and Quinn, *Encyclopedia of Endangered Species*, xiii–xv.

108. Ibid.

109. Ibid., xv.

110. Holm, Evers, and Sasser, "Nutria in Louisiana," 13.

### Chapter 2. Rat Race

1. Account of Hudson's exploits from Hunter, *Half Moon*, 2–3, 6–11, 93–96.

2. See http://www.seinfeldscripts.com/TheChickenRoaster.htm.

3. These are composited sentiments that I have heard expressed again and again by folks I spoke to during the course of research for this book, and I have composited or made pastiches of their voices for clarity, concision, and effect.

4. B. H. Lopez, *Of Wolves and Men*, 138.

5. Ramsey, "Rats to Riches."

6. Novak, "Traps and Trap Research."

7. McGee, "Use of Furbearers."

8. Wright, "Archaeological Evidence."

9. Cited in Müller-Schwarze and Sun, *Beaver*, 136.

10. "Creek Indians," New Georgia Encyclopedia, http://www.georgiaencyclopedia.org/; informational video narrated by Adam Beach, shown to visitors at Casa Grande Ruins National Monument in Coolidge, Arizona.

11. Spence, *Myths of the North American Indians*, 134; Walker, "Sun Dance," 114; Wissler, "Societies of the Plains Indians," 530.

12. C. Martin, *Keepers of the Game*, 63.

13. Cronon, *Changes in the Land*, 34–54.

14. In the sixteenth century, fur had become scarce in Europe after marten pelts were *en vogue*; during the thirteenth and fourteenth centuries, squirrel was the fur of choice; ibid., 99.

15. European and Asian trade in felts and fur can be traced back almost a millennium. Russia, Scandinavia, and central Asia were major suppliers through the fifteenth century, and furs were supplied to the Mediterranean and Middle East through Constantinople. In the ninth and tenth centuries, Scandinavian and Viking traders exported myriad furs to northern and central Europe, including but not limited to marten, reindeer, bear, otter, sable, ermine, black and white fox, and beaver. European beavers were also hunted throughout northern Europe and Siberia, until they became scarce in the seventeenth century as a result of overhunting; see, for example, Fisher, *Russian Fur Trade*, 29, 118–22; McGee, "Use of Furbearers."

16. Until the end of the seventeenth century, beaver hat manufacturers were dependent on the European beaver. The influx of beaver furs from the New World at the end of the seventeenth century increased the number of beaver hats that were made. The highest quality and most expensive hats were made exclusively from beaver wool, also referred to as *castors*. But the price of beaver hats decreased after the production of demi-castors, or half-beavers, that were mixed with wool or hare fur to produce a hat that was of lower quality, but less expensive and similar in style to hats made only from beaver fur. The production of demi-castors was facilitated by *carroting*, which made hare fur felt more easily after the application of mercury nitrate. The constant exposure of hatters to the toxic mercury, and the irregular behavior or dementia that often came with it, led to the phrase *mad as a hatter*; S. Johnson, *Dictionary of the English Language*, 289.

17. Note that the Cajuns' ancestors (Acadians who settled in what are now the Maritime Provinces of Canada) were indentured servants sent to the New World to trap furbearing animals for French consumption. Interestingly, some of the Cajun descendants of these Acadian fur trappers ended up trapping again in southern Louisiana after their expulsion by the British military and subsequent wandering in search of a new homeland.

18. The area that is now Manhattan was long inhabited by the Lenape Indians. In 1524, Florentine explorer Giovanni da Verrazzano sailed under the auspice of King Francis I of France and became the first European to explore the area that eventually became New York City. Verrazzano sailed far enough into the harbor to spot the Hudson River and he named Upper New York Bay the Bay of Santa Margarita—after Marguerite de Navarre—the elder sister of the king; Schwartz, *Mismapping of America*, 42.

19. Cited in Müller-Schwarze and Sun, *Beaver*, 136.

20. According to a letter by Pieter Janszoon Schagen, Peter Minuit and Dutch colonists acquired Manhattan in 1626 from unnamed American Indians, probably Canarsee Indians of the Lenape, in exchange for trade goods worth 60 guilders, often said to be worth US$24, though by some calculations this actually amounts to around US$1,050 in 2014. According to the writer Nathaniel Benchley, Minuit conducted the transaction with Seyseys, chief of the Canarsees, who was willing to accept merchandise in exchange because the island was mostly controlled by the Weckquaesgeeks; Benchley, "$24 Swindle."

21. With a fleet of 21 ships, the Dutch Republic regained New York in August 1673 and renamed the city "New Orange." New Netherland was ceded permanently to the English in November 1674 through the Treaty of Westminster in exchange for Run Island, which was the long-coveted last link in the Dutch nutmeg trading monopoly in Indonesia; Scheltema and Westerhuijs, *Exploring Historic Dutch New York*.

22. J. T. Adams, *Founding of New England*, 102.

23. Obbard et al., "Furbearer Harvests."

24. The notion that increased competition between the English and the French led to overexploitation of beaver stocks by aboriginals is widely accepted, with many historians believing that aboriginals were the primary culprits in depleting animal stocks. However, the effects of historical beaver population dynamics on declining beaver populations remains unclear. Some historians dispute the hypothesis that British-French competition was the primary reason for overexploitation of beaver stocks; other historians focus chiefly on implicating changing economic incentives for aboriginals. According to one study, beaver populations decreased dramatically before the rivalries between European powers in the 1700s, and North American beaver stocks were voraciously harvested before competition between the English and the French; see, for example, Innis, *Fur Trade in Canada*, 9–57; Ray, *Indians in the Fur Trade*, 3–231.

25. Thomson, *Works of James Thomson*, 191.

26. Cited in Müller-Schwarze and Sun, *Beaver*, 140.

27. Ibid.

28. La Salle's idea would not come to fruition in his lifetime. He was murdered by his mutinous crew at age 43 in what is now eastern Texas; Tucker, "La Salle Lands."

29. Arthur, *Fur Animals of Louisiana*.

30. Obbard et al., "Furbearer Harvests."

31. *First Nations* is the term for aboriginal peoples in Canada, analogous to *Native Americans* in the United States.

32. E. E. Rich, *History of the Hudson Bay Company*.

33. These wars are named after the sitting British monarch. As there had already been a King George's War in the 1740s, British colonists named the second war in King George's reign after their opponents, and it became known as the French and Indian War. This traditional name continues as the standard in the United States, but it obscures the fact that Indians fought on both sides of the conflict, and that this was part of the Seven Years' War, a much larger conflict between France and Great Britain. American historians generally use the traditional name or sometimes the Seven Years' War; F. Anderson, *Crucible of War*, 747.

34. The name *French and Indian War* is used mainly in the United States because it refers to the two main enemies of the British colonists: the royal French forces and the various indigenous forces allied with them. British and European historians use the term *Seven Years' War*, as do English-speaking Canadians; Cave, *French and Indian War*, 82.

35. Walton and Rockoff, *History of the American Economy*, 27.

36. Cave, *French and Indian War*, xii.

37. Although part of "French Canada," Montreal was English speaking at this time.

38. Newman, "Canada's Fur Trading Empire."

39. One example among many was a post at Albany Fort by the HBC, which listed that one prime-quality adult beaver skin could be exchanged for one of the following: three-quarters of a pound of colored beads, one brass kettle, one and a half pounds of gunpowder, two pounds of sugar, two pounds of Brazilian tobacco, one gallon of brandy, one blanket, or 12 dozen buttons; Müller-Schwarze and Sun, *Beaver*, 144.

40. Obbard et al., "Furbearer Harvests," 1007.

41. Ray, "Fur Trade in North America."

42. W. R. Smith, *Brief History of the Louisiana Territory*, 36.

43. Cronon, *Changes in the Land*, 107.

44. Sunder, *Fur Trade on the Upper Missouri*, 25–30.

45. In 1822, Henry and William Henry Ashley formed the Rocky Mountain Fur Company; Gilbert, *Westering Man*, 108.

46. Irving, *Adventures of Captain Bonneville*, 27. Not all accounts of the mountain men are romantic. For instance, according to George Frederick Ruxton (ca. 1846), "I have met honest mountain men. Their animal qualities, however, are undeniable . . . they are just what uncivilized white man might be supposed to be in a brute state, depending upon his instinct for the support of life"; cited in Dolin, *Fur, Fortune, and Empire*, 253.

47. Chittenden, *American Fur Trade*, Bison ed., 882–83.

48. All monetary amounts in this book are listed as the value of the money during that year, and are not converted into "today's" dollars unless necessary for reference or effect and explicitly stated in the text; Madsen, *John Jacob Astor*, 45–51.

49. Ibid., 139, 233.

50. An estimate based on inflation from the legally set American gold standard rate of $21 per ounce in the 1850s would result in a much more conservative net worth of $1.272 billion in 2011 dollars; *Life*, December 12, 1960, 16.

51. In 1804, Astor purchased from Vice President Burr what remained of a 99-year lease on property in Manhattan; A. D. H. Smith, *John Jacob Astor*, 65–126, 253–63.

52. Sterba, *Nature Wars*, 73.

53. "The Fur Trade: 'Beaver Powered Engineering,'" available at http://xroads.virginia.edu/~HYPER/HNS/Mtmen/furtrade.html.

54. As with any sheared fur, sheared nutria needs special care and should always be stored in the summer.

55. Lippson and Lippson, *Life in the Chesapeake Bay*, 239.

56. St. Johns Institute for Deaf-Mutes, *Our Young People*, 434.

57. Todd, *Tinkering with Eden*, 199.

58. W. Anderson, "Climates of Opinion"; Osborne, "Acclimatizing the World"; Vavasseur, *Guide du Promeneur*, 11–19.

59. Wallace, *Encyclopaedia Britannica*, 1:114–21.

60. Ibid.

61. Bennett, *Acclimatisation*, 6.

62. The connection between acclimatization societies and nutria presence in Europe is largely speculation. For details on the history of nutria in France, see Carter and Leonard, "Review of the Literature."

63. Dunlap, "Remaking the Land."

64. Palmer, *Dangers of Introducing Noxious Animals*, 87–109; D. S. Smith, "Foreign Birds."

65. E. E. Rich, "Russia and the Colonial Fur Trade."

66. Estlack, *Aleut Internments*, 37–42.

67. Hurtado, *John Sutter*, 49–50.

68. Cited in Manning, *Grassland*, 85.

69. D. Brown, *Bury My Heart at Wounded Knee*.

70. The historical role of buffalo is preserved with various coins, stamps, and the state flag of Wyoming. The same goes for the beaver, now the mascot of Oregon State University, named the national animal of Canada as of 1975, honored in perpetuity with its likeness on the Canadian nickel, and used as a symbol of Canadian agencies such as the Toronto Police Services, Canadian Pacific Railway Police Service, Canadian Pacific Railway, and the cap badges of the Royal 22e Régiment and the Canadian Military Engineers. S. White, "Beaver as National Symbol."

71. Sterba, *Nature Wars*, 73.

72. Ramsey, "Rats to Riches."

73. Lowery, *Mammals of Louisiana*, 137–222.

74. Boscareno, "Rise and Fall of the Louisiana Muskrat," 18–30.

75. O'Neil, *Muskrat in the Louisiana Coastal Marshes*, 46.

76. Lowery, *Mammals of Louisiana*, 137–222.

77. Vaughn, *Social History of the American Alligator*, 63–67.

78. McIlhenny, *Alligator's Life History*, 79.

79. This technique was used historically; now that hydraulic modification has likely decreased the amount and quality of habitat for muskrats and little demand for muskrat fur exists, Louisiana marshes are no longer burned with the express purpose of increasing muskrat populations; Lowery, *Mammals of Louisiana*, 137–222; O'Neil, *Muskrat in the Louisiana Coastal Marshes*, 46.

80. Cited in Holm, Evers, and Sasser, "Nutria in Louisiana," 7.

81. Lowery, *Mammals of Louisiana*, 137–222.

82. Tarver, Linscombe, and Kinler, *Fur Animals*, 1–77.

83. Gomez, *Wetland Biography*, 150.

84. Cited in Baroch et al., "Nutria (*Myocastor coypus*) in Louisiana," 92. The Louisiana Department of Conservation was established in 1910. It was briefly renamed the Louisiana Conservation Commission in 1912 before reclaiming its original name in 1918. In 1944, responsibilities involving wildlife resources were transferred to a new agency called the Louisiana Department of Wildlife and Fisheries, which was renamed the Louisiana Wildlife and Fisheries Commission in 1952. In 1975, the name was changed back to Louisiana Department of Wildlife and Fisheries, which is the current name of the agency; http://www.wlf.louisiana.gov.

85. Ramsey, "Rats to Riches."

86. Linscombe, *1999–2000 Annual Report*, 1–35.

87. Dyhouse, "Skin Deep."

88. Hansen, *Costume Cavalcade*, 1–160.

89. Chase, *Complete Book of Oscar Fashion*, 18, 25, 121.

90. Todd, *Tinkering with Eden*, 199–200.

91. Gomez, *Wetland Biography*, 150.

92. Lowery, *Mammals of Louisiana*, 98–143.

93. Washburn, "Evolution of the Trapping Industry"; Washburn, "Trapper Calls It a Bad Day."

94. Ramsey, "Rats to Riches."

95. Washburn, "Evolution of the Trapping Industry"; Washburn, "Trapper Calls It a Bad Day."

96. Todd, *Tinkering with Eden*, 206.

97. Thrapp, *Encyclopedia of Frontier Biography*, 515.

98. St. Johns Institute for Deaf-Mutes, *Our Young People*, 434.

99. Thrapp, *Encyclopedia of Frontier Biography*, 515.

100. Ibid.

101. Schitoskey, Evans, and LaVoie, "Status and Control of Nutria."

102. Todd, *Tinkering with Eden*, 206.

## Chapter 3. Nutria Gone Wild

1. This is a composite quote from the two letters; both quotes are cited in Bernard, "M'sieu Ned's Rat?," 285, 290–91.

2. Note that McIlhenny Company renders its own name without the preceding article *the*, except when *McIlhenny Company* is used as an adjective; S. K. Bernard, personal communication with the author, November 12, 2015.

3. All quotes in this section from Shane K. Bernard come from the author's personal communication with S. K. Bernard, October 15, 2015. Some of Bernard's quotes are composited or reordered from their occurrence in the discussion for clarity, concision, and effect.

4. Cited in Bernard, "M'sieu Ned's a Rat?," 283.

5. Bernard, "Avery Island," 1. There was just one Bank of Louisiana branch in Baton Rouge, although there were other branches elsewhere; S. K. Bernard, personal communication, November 12, 2015. Edmund's employer, the Bank of Louisiana, used Daniel Avery as one of its attorneys, and it was in this capacity that Edmund met Avery and his family; S. K. Bernard, personal communication, March 16, 2016.

6. Bernard, "Avery Island," 1.

7. Marsh stayed in Baton Rouge to look for a permanent area to settle, and may have visited Petite Anse after reading a description of its bounty in the *Immigrant's Guide to the Western and Southwestern States and Territories* by William Darby. According to the text, Petite Anse was a rich landscape that was good for growing sugar and other plant life, with boundless natural resources. Marsh agreed with the text upon arrival, and during the next 20 years, he purchased most of the southern portion of Petite Anse; Bernard, "Avery Island," 1; G. C. Taylor, "Saga of Petit [*sic*] Anse Island."

8. G. C. Taylor, "Saga of Petit [*sic*] Anse Island."

9. Bernard, *Tabasco*, 29.

10. Ibid., 30.

11. Confederates Major General Richard Taylor, commander of Confederate forces west of the Mississippi, wrote about the importance of the Island's salt in his memoirs: "The want of salt was severely felt in the Confederacy. . . . Intelligence of [Avery's discovery] reached me at New Iberia and induced me to visit the Island. The salt was from 15 to 20 feet below the surface and the underlying soil was soft and friable [crumbly]. Devoted to our cause, Judge Avery placed his mine at my disposition for the use of the government. Many Negroes were assembled to get out salt and a packing establishment was organized at New Iberia to cure beef. During succeeding months large quantities of salt, beef, sugar and molasses were transported by steamers to Vicksburg, Port Hudson and other points East of the Mississippi. Two companies of infantry and a section of artillery were posted on the island to preserve order among the workmen and secure it against a sudden raid of the enemy"; Major G. R. Taylor, *Deconstruction and Reconstruction*, 145.

12. Bernard, *Tabasco*, 30.

13. The Averys and McIlhennys returned to the island after the war. In 1869, D. D. Avery bought out the last of the early landholders. This made D. D. Avery the island's sole owner. When D. D. Avery died in 1879, ownership passed in part to Mary Eliza Avery McIlhenny, D. D. Avery's daughter. Ownership also passed to D. D. Avery's four other children: state militia leader Dudley Avery, Louisiana legislator John Marsh Avery, Sarah Marsh Avery Leeds (wife of Paul B. Leeds, whose family ran the Leeds Foundry in New Orleans), and Margaret Henshaw Avery Johnston (whose husband, soldier and poet William Preston Johnston, served as president of Louisiana State University and first president of Tulane University). In 1903, the surviving Avery siblings organized their plantation into a modern corporation, called the Avery Planting and Improvement Company, an entity that became the Petit [*sic*] Anse Company in 1924 and then Avery Island, Inc., in 1948. Avery Island, Inc., continues to control the island's surface and mineral rights. The island was renamed after D. D. Avery in the late nineteenth century, and present-day shareholders are required by corporate charter to be his direct descendants; Bernard, "Avery Island," 1; S. K. Bernard, personal communication, October 15, 2015. Note that *Petite Anse* is spelled with an *e* on the end of *Petite* because *Anse* is a feminine French word, except when referring to the Petit Anse Company, which always misspelled the word in its title.

14. Although *Tabasco* is the name of a region in Mexico, the origin of the peppers is apparently unknown. Indeed, when Edmund McIlhenny's son John went to Mexico (including the state of Tabasco) in 1890 to search for peppers to augment the island's crop, he could find none that matched those grown by Edmund, rendering the entire matter of the peppers' origin as a mystery. Edmund chose the name *Tabasco* for his product, however, because he liked the sound of the word, and no doubt because the Tabasco region of Mexico was already associated with peppers and other spices. But his choice of name did not necessarily denote that the peppers came from Tabasco, Mexico; S. K. Bernard, personal communication, March 16, 2016.

15. Bernard, *Tabasco*, 31.

16. On his death, Edmund was worth only about $14,000 as the manufacturer of Tabasco sauce; S. K. Bernard, personal communication, November 12, 2015.

17. Bernard, *Tabasco*, 31–32. In most accounts, Edmund obtained the peppers in New Orleans from a soldier, perhaps named "Gleason," who had recently returned to the United States from Mexico, planting the seeds on Avery Island and sharing his homemade sauce in cologne bottles with friends who urged him to sell it in local grocery stores. Perhaps Edmund saw crates of the peppers arriving from the Tabasco region of Mexico and this inspired the saucy idea. How Edmund arrived at the idea to make the sauce also remains unclear. He may have been thinking about a condiment market that had recently become lucrative and realized the niche market for a spicy seasoning. Some folks claim that Edmund invented pepper sauce, others give credit to New Orleans businessman and politician Maunsel White, and others point to references of pepper sauce that predate White's. Neither McIlhenny Company nor the McIlhenny family has ever made the claim that Edmund invented pepper sauce, as they are aware of the prior existence of other pepper sauces like Maunsel White's. Edmund did, however, invent the concept of mass-producing a pepper sauce that through marketing would become a nationally adored product; S. K. Bernard, personal communication, October 15, 2015, and March 16, 2016; S. K. Bernard, personal communication, November 12, 2015. But see R. M. Grace, "Was Col. Maunsel White the True Originator?"

18. Bernard, "Avery Island," 1.

19. With regard to the Tabasco sauce business, John and Ned expanded and modernized the manufacturing process for the sauce, helping their father's invention become commonplace on tables worldwide. In particular it was Edward (Ned), the second son of Edmund and Mary and he of the debate regarding the nutria releases, who experimented with new advertisement methods for Tabasco sauce, such as advertising on radio, full-color posters, and limerick contests. His impact on the company was historic—Ned fought trademark infringements, placed Tabasco promotions in national publications like *Good Housekeeping*, hired demonstrators to teach uses for Tabasco sauce to the public, distributed recipe booklets, and introduced the immediately recognizable screw-top sauce bottle, tweaking the iconic Tabasco diamond logo trademark to create the version seen today; Bernard, *Tabasco*, 115–16, 120.

20. Bernard, *Tabasco*, 99.

21. McIlhenny, *Alligator's Life History*, 7–9. Given Ned's reputation for telling tall tales, it remains unclear as to whether this story is wholly accurate or merely hyperbole.

22. Ned's journals describing Eskimo ways of life were later deposited in the University of Pennsylvania Museum of Archaeology and Anthropology; Deutsch, "Bird That's Not," 1–3; Dickson, "Promotion of Bird City," 1–3.

23. Cited in Bernard, *Tabasco*, 100.

24. Ned's efforts to save the snowy egret kicked into full gear with his assistance to a newsreel company that shot an eight-minute silent movie showing hunters stalking, killing, and dismembering egrets that shocked audiences at theaters, museums, and ecological conferences. Cited in Bernard, *Tabasco*, 100.

25. Bernard, *Tabasco*, 100; Dickson, "Promotion of Bird City," 1–3; Oakes, "Isle of Spice."

26. Cited in Bernard, *Tabasco*, 104. The population of egrets that returned to Avery Island every year eventually reached several hundred thousand, as the area became home to more than 90 percent of the egrets in North America because it was the only place where the birds were safe from hunters. Ned responded to the influx of egrets by expanding the man-made

pond and building a deck to serve as an observation point for the nesting ground, and the refuge remains successful to this day. With the help of businessman and conservationist Charles Willis Ward, the Rockefeller Foundation, and the Sage Foundation, Ned also secured nearly 175,000 acres (710 square kilometers) of southern Louisiana marshland to serve as refuge for wildfowl species like the snowy egret. See also Dickson, "Promotion of Bird City," 1–3; Oakes, "Isle Of Spice."

27. This story is apocryphal; S. K. Bernard, personal communication, November 12, 2015.

28. Bernard, *Tabasco*, 135.

29. Ibid., 123.

30. In fact, the diverse array of flora features some of the most exotic organisms in existence—papyrus from the Nile Valley, Tonkin cane from Southeast Asia, Tibetan podocarp, bamboo and a Wasi orange tree from Japan, finger bananas from China, Indian soap trees, one thousand varieties of camellias—and many others that are hybrids propagated by Ned through years of experimentation. Opened to the public by Ned in 1935 to promote automobile tourism in Louisiana, the gardens eventually offered a landscape architectural service, which designed the grounds of the Louisiana state capitol, Louisiana State University, and other public works projects; Bernard, *Tabasco*, 139.

31. Bernard, "M'sieu Ned's Rat?," 284.

32. This is a composited quote from oral legend that shows how Ned McIlhenny probably told the story to folks on Avery Island; S. K. Bernard, personal communication, October 15, 2015.

33. While it is untrue that Jack London was among the rescued men, it is factual that Ned McIlhenny helped to rescue a large number of whaling fleet sailors. According to Bernard, multiple eyewitnesses and other sources back up the basic story about rescuing the sailors; S. K. Bernard, personal communication, November 13, 2015.

34. S. K. Bernard, personal communication, October 15, 2015.

35. Cited in Bernard, "M'sieu Ned's Rat?," 284.

36. A. P. Daspit, Department of Conservation, to E. A. McIlhenny, October 16, 1930, E. A. McIlhenny Collection, Avery Island, Louisiana; cited in Bernard, "M'sieu Ned's Rat?," 286.

37. E. A. McIlhenny to A. P. Daspit, Department of Conservation, October 18, 1930, E. A. McIlhenny Collection, Avery Island, Louisiana; cited in Bernard, "M'sieu Ned's Rat?," 286.

38. Cited in Bernard, "M'sieu Ned's Rat?," 286.

39. Cited in ibid.

40. Bernard, "Nutria Nuisance," 42.

41. Bernard, "M'sieu Ned's Rat?," 287.

42. W. B. Grant to J. Dymond, president of Delta Duck Club, October 16, 1926, E. A. McIlhenny Collection, Avery Island, Louisiana; cited in Boscareno, "Rise and Fall," 52–54.

43. S. C. Arthur to J. Dymond, president of Delta Duck Club, October 23, 1926, E. A. McIlhenny Collection, Avery Island, Louisiana; cited in Boscareno, "Rise and Fall," 52–54.

44. Author unknown (E. A. McIlhenny?) to S. C. Arthur, Department of Conservation, October 29, 1926, E. A. McIlhenny Collection, Avery Island, Louisiana; cited in Boscareno, "Rise and Fall," 52–53.

45. W. C. Henderson, acting chief of the Bureau of Biological Survey, to W. B. Grant, November 17, 1926, E. A. McIlhenny Collection, Avery Island, Louisiana; cited in Boscareno, "Rise and Fall," 53.

46. E. W. Nelson, chief of the Bureau of Biological Survey, to J. Dymond, president of Delta Duck Club, December 14, 1926, E. A. McIlhenny Collection, Avery Island, Louisiana; cited in Boscareno, "Rise and Fall," 54.

47. Bernard, "M'sieu Ned's Rat?," 287.

48. Carr, "Nutria Tales."

49. Brote, "Cargo Loaded Northbound."

50. Brote, "From: Lieut. Comdr. Henry Conrad Brote."

51. Susan Brote (Mrs. Henry Contrad Brote), Covington, Louisiana, to E. A. McIlhenny (Avery Island, Louisiana), December 16, 1945; cited in Bernard, "M'sieu Ned's Rat?," 287.

52. On the website of Delta World Tire in New Iberia, Louisiana, the "About Us" section of the website refers to the founder of the tire company as "Abe Bernstein." An employee told Bernard that Delta was owned by the descendants of Mr. Bernstein, the fur agent. Bernard also recently saw a photo of downtown New Iberia circa the 1920s that showed a sign for Mr. Bernstein with the phrase "fur trapper" (or "fur company") under it. Thus, Abe Bernstein is almost certainly the "A. Bernstein" mentioned in McIlhenny's papers. S. Bernard, personal communication, July 27, 2015; also see http://www.deltaworldtire.com/.

53. Cited in Bernard, "M'sieu Ned's Rat?," 287.

54. Cited in ibid., 288.

55. Cited in ibid.

56. Cited in ibid.

57. Cited in ibid., 289.

58. Ibid.

59. Bernard, "M'sieu Ned's Rat?," 288–89.

60. "Louisiana Declares War," 1–3.

61. Bernard, "M'sieu Ned's Rat?," 290.

62. Ibid.

63. Ibid.

64. Bernard, "M'sieu Ned's Rat?," 290–91.

65. Bernard, *Tabasco*, 132.

66. Ibid., 140.

67. S. K. Bernard, personal communication, July 27, 2015.

68. Bernard, *Tabasco*, 142.

69. Bernard, "Avery Island," 1.


## Chapter 4. Alien Invasion

1. See http://www.seinfeldscripts.com/TheReversePeephole.htm.

2. After government officials ignored several reports encouraging camel-based transport, Senator Jefferson Davis of Mississippi came across the idea and was riveted. When Davis was appointed secretary of war during a period of westward exploration in 1853, he campaigned to use camels for military purposes and eventually gained congressional support. By 1855, the US Congress had appropriated funds for the project. To acquire camels and drivers, outspoken supporter Major Henry C. Wayne was dispatched on board the USS

*Supply.* Wayne traveled throughout the coasts of the Mediterranean Sea, and after obtaining 33 camels (or close evolutionary relatives) and five drivers, the USS *Supply* sailed to coastal Texas. Camels were transported to Camp Verde by May 1856, and the ship turned back for more, eventually growing the camel herd to 70 in 1857; Manno, "Tedro's Last Stand."

3. Ibid. Hadji Ali was an Ottoman citizen born in modern-day Jordan. Born of Greek and Syrian ancestry, Ali was a camel breeder and trainer for the Ottoman military, and had also served in Algiers for the French army. His charisma helped him sign on with the Camel Corps in 1856 and entertained his American masters, who anglicized his name to Hi Jolly.

4. Ibid.

5. Mirsky, "Antigravity."

6. The pigs were released into the wild by wealthy landowners and were meant to serve as big game animals. Initial introductions took place in fenced enclosures and several pigs escaped, with the escapees sometimes interbreeding with already established feral pig populations. The first introduction occurred in New Hampshire during 1890, with some individuals crossing into Vermont. In 1902, wild boars from Germany were released into an estate in upstate New York. The most successful boar introduction in the United States took place in western North Carolina in 1912, when 13 boars from Europe were released into two fenced enclosures in a game preserve in Hooper Bald, Graham County. A large-scale hunt catalyzed the escapes of these individuals, who were later found in Tennessee, where they interbred with other pigs in the area. Other escapes under similar circumstances in California and Texas led to widespread boar populations; Mayer, *Wild Pigs.*

7. Kudzu is a legume that increases the nitrogen in the soil via a symbiotic relationship with nitrogen-fixing bacteria. Its deep taproots also transfer valuable minerals from the subsoil to the topsoil, thereby improving the topsoil; Cain, Bowman, and Hacker, *Ecology*, 246.

8. Canover, Simmonds, and Whalen, *Management and Control Plan*, 21–27.

9. Gomez, "Perspective, Power, and Priorities"; "Commercial Value of 1928–29 Fur Crop."

10. "Muskrat Changes Its Name"; Branan, "Muskrat vs. Mink," 61.

11. The Louisiana Department of Conservation was originally established in 1910. The Department of Conservation was briefly renamed the Louisiana Conservation Commission in 1912 before reclaiming its original name in 1918. State wildlife responsibilities were transferred to a new agency called the Louisiana Department of Wildlife and Fisheries in 1944, which was also briefly renamed the Louisiana Wildlife and Fisheries Commission in 1952. That name applied until 1975, when the name was changed back to the current Louisiana Department of Wildlife and Fisheries.

12. McHugh, "Conservation Second Only to War."

13. Recipes included fried muskrat, muskrat pie, and muskrat with tomato sauce. "Louisiana Muskrat May Help Alleviate"; "Fats from Fur Animals," 5; "Dignitaries Give Louisiana Muskrat Approval," 7; Dozier, "Suggested Muskrat Recipes," 4; Gowanloch, "Louisiana's Muskrat Industry."

14. O'Neil, "Fur Luxury."

15. O'Neil, *Muskrat in the Louisiana Coastal Marshes*, 1–28.

16. Ibid.

17. Chabreck, Linscombe, and Kinler, "Winter Food."

18. Daspit, "Trappers Pile Up," 16.

19. Washburn, "Plan Fur Industry Comeback," 12.

20. Ashbrook, "Muskrat-Nutria"; Svihla and Svihla, "Louisiana Muskrat."

21. Daspit, "Trappers Pile Up," 16; Washburn, "Trapper Calls It," 12.

22. Ashbrook, "Nutrias Grow"; Morgan and O'Neil, "Stalking the Wild Fur."

23. Evans, "Nutria." Hurricanes were not named until 1953, when the National Weather Service began giving them female names. Male and female names were used starting in 1978; see "Tropical Cyclone Naming History and Retired Names," http://www.nhc.noaa.gov/aboutnames_history.shtml.

24. Ashbrook, "Nutrias Grow."

25. Evans, "Nutria"; O'Neil and Linscombe, "Fur Animals." Nutria were well established throughout the coastal areas of Louisiana by 1943, and exhibited rapid population growth for a number of years thereafter. Indications of nutria population levels in Louisiana since 1943 are largely indirect, and come from two sources: (1) annual pelt harvest records as reflected in state severance tax records; and (2) incidence and degree of nutria damage to crops, levee systems, and native marsh habitats. Local nutria population levels have also been estimated using direct methods such as mark-recapture, night counts, and indirect indexes such as scat counts and active trail counts, but historically, the primary indictor of the statewide nutria population has been the annual fur-trapping harvest level derived from severance tax records; Robicheaux and Linscombe, "Effectiveness of Live-Traps"; Spiller and Chabreck, "Wildlife Populations."

26. O'Neil, "Fur Industry in Retrospect"; "Last Year's Fur Crop."

27. Ashbrook, "Nutrias Grow."

28. "Nutria Added to List."

29. The Louisiana Department of Wildlife and Fisheries now classifies 13 of the state's mammals as "official" furbearers; along with the nutria, the beaver and muskrat are the other two rodents. The five furbearers that have shaped the history of Louisiana's fur history the most are the nutria, muskrat, mink, otter, and raccoon; Gomez, *Wetland Biography*, 143.

30. Gomez, *Wetland Biography*, 151–52.

31. Cited in ibid., 151.

32. V. T. Harris, "Nutria as a Wild Fur Mammal"; Evans, "Nutria."

33. O'Neil, "From Eight Million Muskrats."

34. In 1961–62, nutria surpassed muskrat in both total harvest and value; O'Neil, "Louisiana's Fur Industry Today."

35. Daspit, "'Migrating' Nutria."

36. Kinler et al., "Effect of Tidal Flooding."

37. Cited in Gomez, *Wetland Biography*, 151.

38. Boscareno, "Rise and Fall," 1.

39. During the peak muskrat populations, prolonged wet cycles accompanied mild temperatures; O'Neil, "From Eight Millions Muskrats."

40. This is an astounding number, considering that Ned McIlhenny's original captive population was 20 animals; O'Neil and Linscombe, "Fur Animals."

41. O'Neil, "Louisiana's Fur Industry Today," 5.

42. Harris and Chabreck, "Some Effects of Hurricane Audrey"; Webert, "Hurricane Damage to Fur Industry."

43. Harris and Chabreck, "Some Effects of Hurricane Audrey"; Webert, "Hurricane Damage to Fur Industry." So great was the damage and loss of life that the name Audrey was later retired from usage as an identifier for an Atlantic hurricane, which is sometimes done by the National Weather Service because of sensitivity issues with using the name again; Ross and Blum, "Hurricane Audrey 1957."

44. Harris and Chabreck, "Some Effects of Hurricane Audrey"; Waldo, "Hurricane Damages Refuges."

45. Ensminger, "Economic Status of Nutria."

46. Ibid.; Evans, "Nutria."

47. Ensminger, "Economic Status of Nutria"; Evans, "Nutria."

48. Mouton, Linscombe, and Hartley, *Survey of Nutria Herbivory Damage*, 1–10.

49. Obbard et al., "Furbearer Harvests."

50. Ibid.

51. Ibid.

52. Edmond Mouton, LDWF, personal communication, cited in Holm, Evers, and Sasser, "Nutria in Louisiana," 6.

53. Holm, Evers, and Sasser, "Nutria in Louisiana," 5.

54. A little over a decade later, Jean-Baptiste Le Moyne de Bienville, the first administrator of France's Louisiana colony, succeeded in growing sugarcane brought from Martinique in his garden at New Orleans; Gayarré, *History of Louisiana*, 2:62–70.

55. Ibid.

56. Ibid.

57. Ibid.; Abbott, *Sugar*, 270–310.

58. Abbott, *Sugar*, 270–310.

59. Chenrow and Chenrow, *Reading Exercises in Black History*, 52.

60. Prichard, "Effects of the Civil War."

61. Evans, "Nutria."

62. Holm, Evers, and Sasser, "Nutria in Louisiana," 5.

63. Ibid., 6.

64. Kinler, Linscombe, and Ramsey, "Nutria."

65. The practice of using nutria teeth, also known as "swamp ivory," for jewelry continued sparingly into the 1990s, most notably with a woman in Washington, Louisiana, named June Lowery, who started a small business after acquiring 100 nutria heads and consulting local dentists about tools. The business ended abruptly when she realized that nutria teeth become brittle with age; Trillin, "Nutria Problem."

66. Tarver, Linscombe, and Kinler, *Fur Animals*, 1–6.

67. Emberly, *Cultural Politics of Fur*, 1–2, 36.

68. Ibid., 1.

69. In the same year, Edward Abbey published his most famous fiction work, *The Monkey Wrench Gang* (New York: Harper Collins, 1975), which discusses using sabotage to protest

environmentally damaging activities in the American Southwest, and was highly influential in the pro-environmental movement.

70. Misiroglu, *American Counterculture*, 552.

71. Stone, "Foreword," in Guillermo, *Monkey Business*.

72. Cited by Garner, *Political Theory of Animal Rights*, 144.

73. Ibid.; Phelps, *Longest Struggle*, 242.

74. Abel, *Aboriginal Resource Use*, 329.

75. Phelps, *Longest Struggle*, 242.

76. Baroch et al., "Nutria in Louisiana," 93.

77. Ibid., 92.

78. Emberly, *Cultural Politics of Fur*, 2.

79. Cited in Baroch et al., "Nutria in Louisiana," 92.

80. Not all celebrities are against fur: Cindy Crawford and Jennifer Lopez have both been criticized for wearing fur products to public events; Freeman, "Fighting the Return of Fur."

81. Wiebe and Mouton, "Nutria Harvest and Distribution."

82. Mouton, Linscombe, and Hartley, *Survey of Nutria Herbivory Damage*, 1–10.

83. Although damage to wetlands became the primary focus in the 1990s, nutria were estimated to have caused almost $2 million in damage to sugarcane. and also continued to cause damage to rice levees. Currently the damage to agricultural crops by nutria is significant but not as salient as during the conflict between the sugar and fur industries in the 1950s and '60s; Hebert, "Agricultural Damage," 1–7.

84. Mouton, Linscombe, and Hartley, *Survey of Nutria Herbivory Damage*, 1–10.

85. Cited in Baroch et al., "Nutria in Louisiana," 29.

### Chapter 5. Damage Control

1. T-Boy is a symbolic creation to commemorate Groundhog Day each year. He (she?) changes each year and often is acquired (to ensure he is the right size) specifically for the fun event during which he pops up (from inside the Superdome, a Mardi Gras float, etc.). T-Boy has no exhibit of his own. The nutria brought in is then placed in the Swamp Exhibit with the other nutria. One would think that these disclaimers might also apply to other weather-predicting nutria. For weather reports from T-Boy, see Jing, "Forget Phil!," 1; Saul, "T-Boy the Nutria," 1.

2. Rurik, "Shadeaux Sees Neaux Shadow," 1–3.

3. Website of the New Orleans Zephyrs: http://www.milb.com/index.jsp?sid=t588.

4. Penland and Ramsey, "Relative Sea-Level Rise."

5. Barras, Bourgeois, and Handley, *Land Loss Rates*.

6. Galluci, "BP Oil Spill."

7. Valentine et al., "Alligator Diets."

8. Sheng, Lapetina, and Ma, "Reduction of Storm Surge."

9. Gersberg et al., "Role of Aquatic Plants."

10. The major study on this topic places the subsiding of land along Louisiana's Mississippi River delta at 4–11 millimeters per year; Wanless, "Final Report and Findings," 1–11.

11. Emanuel, "Increasing Destructiveness."

12. These results held for both a severe Katrina-like hurricane and a more modest hurricane, both of which were considered to be making landfall at about 12 mph; Wamsley et al. "Influence of Wetland Degradation."

13. Farber, "Welfare Loss of Wetlands Disintegration."

14. Farber and Costanza, "Economic Value of Wetland Systems."

15. Cited in Louisiana Department of Wildlife and Fisheries, *Nutria: Wetland Damage.*

16. Baroch et al., "Nutria in Louisiana," 110.

17. See, for example, Evers et al., "Impact of Vertebrate Herbivores"; Ford and Grace, "Effects of Vertebrate Herbivores."

18. Schmidt, "Nutria Burrowing."

19. Guichon and Cassini, "Local Determinants of Coypu Distribution."

20. Baroch et al., "Nutria in Louisiana," 21, 44–61.

21. Keddy et al., "Wetlands of Lakes Pontchartrain and Maurepas."

22. Coleman, Roberts, and Stone, "Mississippi River Delta."

23. Keddy et al., "Wetlands of Lakes Pontchartrain and Maurepas."

24. Chabreck, "Vegetation, Water and Soil Characteristics."

25. Ibid.

26. Ibid.

27. S. K. Bernard, personal communication, March 16, 2016.

28. Davis, "Historical Perspective."

29. Woolfolk, *Tangipahoa Crossings.*

30. Heleniak and Dranguet, "Changing Patterns of Human Activity."

31. Block, *Early Sawmill Towns*; Burns, "Frank B. Williams."

32. Norgress, "History of the Cypress Lumber Industry."

33. Burns, "Frank B. Williams."

34. Fricker, *Historic Context.*

35. Stolzenburg, "Swan Song of the Ivory-Bill."

36. Visser et al., "Louisiana Coastal Area."

37. Brewer and Grace, "Plant Community Structure."

38. Prescribed burning is used even though it may also decrease biodiversity if organic matter in the soil is ignited because this creates new depressions and increased flooding; see, for example, Lane et al., "Potential Nitrate Removal"; Vogl, "Effects of Fire."

39. Ford and Grace, "Effects of Vertebrate Herbivores."

40. Gough and Grace, "Effects of Flooding, Salinity."

41. Louisiana Department of Wildlife and Fisheries, *Monitoring Plan: Project No. LA-02,* 1.

42. McFalls et al., "Hurricanes, Floods, Levees, and Nutria."

43. Ibid.

44. Grace and Ford, "Potential Impact of Herbivory."

45. Galluci, "BP Oil Spill," 1–3; Mendelssohn et al., "Oil Impacts on Coastal Wetlands."

46. J. Carter, personal communication, March 22, 2016.

47. Keddy et al., "Alligator Hunters."

48. Bartram, *Travels*, 123.

49. Audubon, "Observations," 271.

50. Keddy et al., "Alligator Hunters."

51. Wolfe, Bradshaw, and Chabreck, "Alligator Feeding Habits."

52. Audubon, "Observations," 271.

53. McIlhenny, *Alligator's Life History*, 76.

54. Joanen and McNease, "Management of Alligators."

55. O'Neil, *Muskrat in the Louisiana Coastal Marshes*, 1–20.

56. Cited in Baroch et al., "Nutria in Louisiana," 47.

57. O'Neil, *Muskrat in the Louisiana Coastal Marshes*, 1–20.

58. The authors based diet composition on actual or estimated live weights of intact, undigested prey items. Earlier studies often relied on occurrence and weight of partially digested remains to estimate the levels of diet components; Wolfe, Bradshaw, and Chabreck, "Alligator Feeding Habits."

59. Valentine et al., "Alligator Diets."

60. Keddy et al., "Alligator Hunters."

61. Ibid.

62. Ibid.

63. The report result is for fresh and intermediate marsh types; McNease et al., "Distribution and Relative Abundance."

64. Visser et al., "Long-Term Vegetation Change."

65. McNease, "Distribution and Relative Abundance."

66. Approximately 14 percent of the 48-inch-long alligators are released each year to augment the wild breeding stock in the marshes. The percentage released is adjusted periodically based on breeding success in the wild, survival rates of different sizes of alligators, and other factors. Drought conditions have led to decreased egg collections in certain years, like 1998, but higher water levels have caused increased egg collections in other years, such as 1999. Hides and meat produced by the alligator-farming industry were valued at over $12 million in 2001, and both the wild and farmed alligator industries are tied to the success of wild alligator populations; Elsey, "Changes in Louisiana's Alligator Management Program."

67. Alligator distribution is restricted to freshwater, intermediate, and brackish marshes, or to mangrove swamps with limited salinity, because they have buccal salt-secreting glands and can therefore tolerate only moderate salinity levels; Louisiana Department of Wildlife and Fisheries, "Louisiana's Alligator Management Program."

68. *A. mississippiensis* was removed from the International Union for the Conservation of Nature (IUCN) "Red List" of threatened species in 1996. The IUCN Crocodile Specialist Group (CSG) action plan for the species currently considers the availability of population survey data to be "good," the need for wild population recovery to be "low," and the potential for sustainable management "high"; cited in Baroch et al., "Nutria in Louisiana," 46–47.

69. Cited in Baroch et al., "Nutria in Louisiana," 45.

70. Ibid.

71. The value was estimated at $4 million in 2001; Linscombe, *1999–2000 Annual Report*.

72. Cited in Holm, Evers, and Sasser, "Nutria in Louisiana," 57.

73. Coastal Wetlands Planning, Protection and Restoration Act website, http://lacoast.gov/new/default.aspx.

74. Keddy et al., "Alligator Hunters."

75. Lake Pontchartrain Basin Foundation, *Comprehensive Management Plan*.

76. Keddy et al., "Alligator Hunters."

77. Ibid.

78. This component of restoration was highlighted in a 2004 conservation area plan for the Lake Pontchartrain Estuary completed by the Nature Conservancy; Nature Conservancy, *Conservation Area Plan*.

79. Keddy et al., "Alligator Hunters."

80. Schleifstein, "Vitter Backs Off"; Capitol News Bureau, "House Votes."

81. Blair and Langlinais, "Nutria and Swamp Rabbits."

82. Conner and Toliver, "Vexar Seedling Protectors."

83. Sheffels et al., "Efficacy of Plastic Mesh Tubes."

84. Blair and Langlinais, "Nutria and Swamp Rabbits"; Conner and Toliver, "Vexar Seedling Protectors"; Myers, Shaffer, and Llewellyn, "Baldcypress Restoration."

85. Fuller et al., "Effects of Herbivory."

86. Sheffels et al., "Efficacy of Plastic Mesh Tubes."

## Chapter 6. Bounty Hunters

1. All quotes in this section are from C. McCree, personal communication, July 1, 2015.

2. "Greg Williams Wanted 49ers Hurt," ESPN.com, April 6, 2012, http://espn.go.com/nfl/story/_/id/7778005/gregg-williams-told-new-orleans-saints-hurt-san-francisco-49ers-speech.

3. Manuel and Mouton, *Coastwide Nutria Control Program*, 4–8. Note that incentive payments are provided only for nutria with tails greater than eight inches long to prevent trappers from cutting tails in half and acquiring multiple bounties from the same tail. Some might think that this "eight-inch rule" is in place to ensure that the tail is large enough to belong to an adult rather than a kit, but given the interests of controlling nutria populations, LDWF does not seem particularly interesting in protecting kits.

4. Terrebonne Parish turned in the most tails: 130,952. Saint Mary and Saint Martin Parishes followed with 58,229 and 54,027, respectively; ibid., 8, 13, 24.

5. Ibid., 20, 41, 47.

6. Jordan and Mouton, *Coastwide Nutria Control Program*, 9.

7. Manuel and Mouton, *Coastwide Nutria Control Program*, 4–8.

8. Donlan and Donlan, "Trappers of the Barataria," 4.

9. Manuel and Mouton, *Coastwide Nutria Control Program*, 12.

10. Link, "Living with Nutria," 4–6.

11. Ibid., 5–6.

12. Ibid.

13. Poché, "Recent Studies with Norway Rats."

14. Baroch et al., "Nutria in Louisiana," 145–47.

15. Link, "Living with Nutria," 6.

16. Hilton and Robinson, "Fate of Zinc Phosphide."

17. LeBlanc, "Nutria."

18. Baroch et al., "Nutria in Louisiana," 119–25.

19. Ibid., 126–27.

20. See, for example, Balser, "Management of Predator Populations"; Travis and Schaible, "Effects of Diethylstilbestrol."

21. Ericsson, "Male Antifertility Compounds."

22. Nutria Management Team, *Chesapeake Bay Eradication Project*, 1–2.

23. Ibid.

24. Bounds, "Marsh Restoration"; Gosling, "Towards an Eradication Plan."

25. Southwick Associates, *Potential Economic Losses*, 1–16.

26. Horton, "Chesapeake's Marshes," 1–3; Fears, "In Maryland, a Renewed Effort," 1–3.

27. Federal Leadership Committee for the Chesapeake Bay, "Strategy for Protecting and Restoring," 6.

28. Fears, "In Maryland, a Renewed Effort," 1–3; Emery, "Killed by the Thousands," 1–3.

29. Emery, "Killed by the Thousands," 1–3.

30. Six, "Biologist Says Nutria Have Arrived," 1–3.

31. J. Carter, personal communication from interview on July 28, 2015.

32. Sheffels and Sytsma, *Report on Nutria Management and Research*, 1–20.

33. Kuhn and Peloquin, "Oregon's Nutria Problem," 101–2.

34. Larrison, "Feral Coypus."

35. Kuhn and Peloquin, "Oregon's Nutria Problem," 101–2.

36. Ibid.

37. Ibid., 101–5.

38. Bernstein, "Nutria Dispute"; Mortenson, "Gresham Dog Dies."

39. Sheffels and Sytsma, *Report on Nutria Management and Research*, 1–20.

40. S. Harris, "Nutria—An Invasive Species," 1–2.

41. Ashbrook, "Nutrias Grow."

42. Larrison, "Feral Coypus."

43. Sheffels and Sytsma, *Report on Nutria Management and Research*, 1–20.

44. Kruse, "Impact of Nutria," 17–18.

45. Schitoskey, Evans, and LaVoie, "Status and Control."

46. Borlick, "NC, VA Seek to Stop."

47. Griffo, "Status of Nutria."

48. "Nutria: *Myocastor coypus*," Florida Fish and Wildlife Conservation Commission, http://myfwc.com/wildlifehabitats/nonnatives/mammals/nutria/.

49. Carter and Leonard, "Review of the Literature."

50. Laurie, "Coypu (*Myocastor coypus*) in Great Britain."

51. Warwick, "Some Escapes of Coypus."

52. Gosling and Baker, "Planning and Monitoring an Attempt."

53. Gosling, Watt, and Baker, "Continuous Retrospective Census."

54. At the time, the Agriculture Department was called the Ministry of Agriculture, Fisheries, and Food; Gosling and Baker, "Eradication of Muskrats and Coypus."

55. Ibid.; "Police to Investigate Man." In 1994, Dr. Morris Gosling, who was a leader in the United Kingdom's anti-nutria initiatives, visited with wildlife officials in Maryland to collaborate on the creation of a comprehensive nutria-eradication program. Correctly predicting that eradication, or at least near-eradication, was an achievable goal, Dr. Gosling suggested that Maryland officials collect information on how nutria behave and reproduce in Maryland's habitats, which were subtly different from what he worked with in England. His suggestion proved useful, and the knowledge gained from the research has helped Maryland come close to ridding the state of nutria. See also Gosling, "Towards an Eradication Plan."

56. J. Carter, personal communication from interview on July 28, 2015.

57. Ibid.

58. Cocchi and Riga, "Nutria *Myocastor coypus*," 38.

59. Ibid.

60. Bertolino and Viterbi, "Long-Term Cost-Effectiveness."

61. Carter and Leonard, "Review of the Literature."

62. Ibid.

63. Ibid.

64. Ibid.

65. Harper, Mavuti, and Muchiri, "Ecology and Management."

66. Carter and Leonard, "Review of the Literature."

67. Ibid.

68. Ibid.

69. J. Carter, personal communication from interview on July 28, 2015.

70. Kanwanich, "Arguments for and against Breeding," 1–2.

71. J. Carter, personal communication from interview on July 28, 2015.

72. Miura, "Dispersal of Nutria."

73. Transliterated from Japanese in I. Shinsuke, Okayama, Japan, Asahi Shimbun, April 14, 1996; cited in Carter and Leonard, "Review of the Literature," 162.

74. Han, Kong, and Cha, *Exotic Animal Species*.

75. Buskey, "In the Market for Nutria?," 1–2.

76. Buskey, "Can Nutria Fur?," 1.

77. Cooper, "Louisiana Is Trying," 1–2.

78. P. Parola, personal communication from interview on October 16, 2015.

79. J. Carter, personal communication from interview on July 28, 2015.

80. Saadoum, Cabrera, and Castellucio, "Fatty Acids, Cholesterol and Protein Content."

81. Cooper, "Louisiana Is Trying," 1–3; A. Ensminger, personal communication, cited in Baroch et al., "Nutria in Louisiana," 48.

82. Evans and Ward, "Secondary Poisoning"; other instances cited in Baroch et al., "Nutria in Louisiana," 48–49.

83. Buskey, "In the Market for Nutria?," 1–2.

84. Cooper, "Louisiana Is Trying," 1–3.

85. Wilkinson, "'Try It—You'll Like It.'"

86. Ibid.

87. Ibid.

88. All quotes in this section are from P. Parola, personal communication from interview on October 16, 2015. Parola's quotes do not necessarily appear in the order in which he stated them, and some are composited for clarity, concision, and effect.

89. Wilkinson, "'Try It—You'll Like It,'" 25.

90. According to Parola, Egon was a native of Romania and spent several years of his childhood as a captive of a Nazi concentration camp during World War II.

91. Cooper, "Louisiana Is Trying," 1–3.

92. Wilkinson, "'Try It—You'll Like It,'" 25.

93. Ibid.

94. Ibid.

95. Ibid.

## Chapter 7. Afterword

1. This sentiment has been expressed by several people in some form or another. One example is Doug Wachob, associate executive director overseeing the Conservation Research Center, advancement, and property management at Teton Science Schools, who used a paraphrase of this idea during a lecture at the Teton Science School in Jackson, Wyoming, on a date that I cannot remember. Another example comes from Steed, "Why Don't We Just Shoot Them?," 2.

2. Emanuel, "Increasing Destructiveness."

3. Panel on Advancing the Science of Climate Change, National Research Council, *Advancing the Science of Climate Change*, 1, 22.

4. Roach, "Is Global Warming Making Hurricanes Worse?," 1–2.

5. Israel, "Swamp Rats on the Move."

6. Ibid.

7. J. Carter, personal communication from interview on July 27, 2015; Le Coz, "Hurricane Isaac Leaves Nutria Dead," 1–3.

8. J. Houge, personal communication from interview on July 1, 2015.

9. Lamb, Reading, and Andelt, "Attitudes and Perceptions."

10. National Trappers Association website, http://www.nationaltrappers.com/.

# Selected Bibliography

Abbas, A. "Feeding Strategy of Coypu." *Journal of Zoology* 224 (1991): 385–401.

Abbott, E. *Sugar: A Bittersweet History*. New York: Overlook Press, 2010.

Abel, K. *Aboriginal Resource Use in Canada*. Winnipeg: University of Manitoba Press, 1991.

Adams, J. T. *The Founding of New England*. Boston: Atlantic Monthly Press, 1921.

Adams, W. H., Jr. "The Nutria in Coastal Louisiana." *Proceedings of the Louisiana Academy of Sciences* 19 (1956): 28–41.

Addison, J. D., ed. *Nutria: Destroying Marshes the Old Fashioned Way*. New Orleans: Louisiana Coastal Wetlands Conservation and Restoration Task Force, 2000.

Alexander-Bloch, B. "Louisiana Commercial Fishery Tops Gulf of Mexico in Total Catch and Revenue." *New Orleans Times-Picayune*, April 29, 2014, http://www.nola.com/environment/index.ssf/2014/04/louisiana_commercial_fishery_t.html.

Aliev, F. F. "Contribution to the Study of Nutria-Migrations, *Myocastor coypus* (Molina, 1782)." *Saugetierkundliche Mitteilungen* 16 (1968): 301–3.

———. "Enemies and Competitors of the Nutria in USSR." *Journal of Mammalogy* 47 (1966): 353–55.

———. "Extent and Causes of Nutria Mortality in the Water Bodies of the Southern USSR." *Mammalia* 29 (1965): 435–37.

———. "Growth and Development of Nutrias' Functional Features." *Fur Trade Journal of Canada* 42 (1965): 2–3.

———. "Numerical Changes and the Population Structure of the Coypu (*Myocastor coypus*, Molina) in Different Countries." *Saugetierkundliche Mitteilungen* 15 (1966): 238–42.

American Heritage Dictionaries. *Spanish Word Histories and Mysteries—English Words That Came from Spanish*. New York: Houghton Mifflin, 2007.

Anderson, F. *Crucible of War: The Seven Years' War and the Fate of Empire in British North America, 1754–1766*. New York: Alfred A. Knopf, 2000.

———. *The War That Made America: A Short History of the French and Indian War*. New York: Viking, 2005.

Anderson, W. "Climates of Opinion: Acclimatization in Nineteenth-Century France and England." *Victorian Studies* 35 (1992): 135–37.

Arthur, S. C., to J. Dymond, president of Delta Duck Club, October 23, 1926, E. A. McIlhenny Collection, Avery Island, LA.

———. *The Fur Animals of Louisiana*, Bulletin 18. Louisiana Department of Conservation. Baton Rouge: Ramires-Jones, 1931.

Ashbrook, F. G. "Muskrat-Nutria." *Louisiana Conservationist* 5 (1953): 16–17.

———. "Nutrias Grow in United States." *Journal of Wildlife Management* 12 (1948): 87–95.

Ashley, William H. *Enterprise and Politics in the Trans-Mississippi West*. Norman: University of Oklahoma Press, 1980.

Atwood, E. L. "Life History Studies of Nutria, or Coypu, in Coastal Louisiana." *Journal of Wildlife Management* 14 (1950): 249–65.

Audubon, J. J. "Observations on the Natural History of the Alligator." In *The Edinburgh New Philosophical Journal, Exhibiting a View of the Progressive Discoveries and Improvements in the Sciences and the Arts*, 270–80. London: A. and C. Black, 1827.

Babero, B. B., and J. W. Lee. "Studies on the Helminths of Nutria, *Myocastor coypus* (Molina), in Louisiana with Check List of Other Worm Parasites from This Host." *Journal of Parasitology* 47 (1961): 378–90.

Balser, D. S. "Management of Predator Populations with Antifertility Agents." *Journal of Wildlife Management* 28 (1964): 352–58.

Bar-Ilan, A., and J. Marder. "Adaptations to Hypercapnic Conditions in the Nutria (*Myocastor coypus*)—In Vivo and In Vitro $CO_2$ Titration Curves." *Comparative Biochemistry and Physiology* 75A (1983): 603–8.

Baroch, J., M. Hafner, T. L. Brown, J. J. Mach, and R. M. Poché. "Nutria (*Myocastor coypus*) in Louisiana." Wellington, CO: Genesis Laboratories (2002): 1–155.

Barras, J. A., P. E. Bourgeois, and L. R. Handley. *Land Loss Rates in Coastal Louisiana, 1956–90*. Lafayette, LA: National Wetlands Research Center Open File Report 94–01, 1994.

Barry, J. M. *Rising Tide: The Great Mississippi Flood of 1927 and How It Changed America*. New York: Simon and Schuster, 1998.

Bartram, W. *Travels through North and South Carolina, Georgia, East and West Florida, the Cherokee Country, the Extensive Territories of the Muscogugles, or Creek Confederacy, and the Country of the Choctaws*. Philadelphia: James and Johnson, 1791.

Bean, M. J. *The Evolution of National Wildlife Law*, 2nd ed. New York: Praeger, 1983.

Beck, E. "The Ecclesiastical Hat in Heraldry and Ornament before the Beginning of the 17th Century." *Burlington Magazine for Connoisseurs* 22 (1913): 338–344.

Beeton, S. O. *Beeton's Brave Tales, Bold Ballads, and Travels and Perils by Land and Sea*. London: Warwick House, 1872.

Benchley, N. "The $24 Swindle: The Indians Who Sold Manhattan Were Bilked, All Right, but They Didn't Mind—The Land Wasn't Theirs Anyway." *American Heritage* 11 (1959): 1.

Bennett, G. *Acclimatisation: Its Eminent Adaptation to Australia*. Melbourne: William Goodhugh.

Bernard, S. K. "Avery Island." *KnowLA Encyclopedia of Louisiana*. Edited by D. Johnson. Louisiana Endowment for the Humanities, September 21, 2012, http://www.knowla.org/entry/1210/&view=article.

———. *The Cajuns: Americanization of a People*. Jackson: University Press of Mississippi, 2003.

———. "M'sieu Ned's Rat? Reconsidering the Origin of Nutria in Louisiana: The E. A. McIlhenny Collection, Avery Island, Louisiana." *Louisiana History: The Journal of the Louisiana Historical Association* 43 (2002): 281–93.

———. "The Nutria Nuisance." *Louisiana Cultural Vistas* (Winter 2011): 42–44.

———. "Soldier, Patriot, Christian, Gentleman: A Biographical Sketch of John Avery McIlhenny." *Attakapas Gazette*, 1993.

———. *Tabasco: An Illustrated History.* Avery Island, LA: McIlhenny Company, 2007.

Bernstein, M. "Nutria Dispute in SE Portland Ends with Dead Dog, Aggravated Animal Abuse Charge." *Oregonian/Oregon Live*, January 24, 2014.

Bertolino, S., and R. Viterbi. "Long-Term Cost-Effectiveness of Coypu (*Myocastor coypus*) Control in Piedmont (Italy). *Biological Invasions* 12 (2009): 2549–58.

Best, M. S. "A Scent Station-Trapping Technique for Developing Furbearer Population Indices in East Texas." Master's thesis, Stephen F. Austin State University, 1988.

Bevington, A. "Arctic Profiles." *Arctic* 36 (1983): 386–87.

Bienvenu, M. "Nutria Goes on the Menu." *Louisiana Cookin'*, May–June 1999, 1–12.

Blair, R. M., and M. J. Langlinais. "Nutria and Swamp Rabbits Damage Baldcypress Plantings." *Journal of Forestry* 58 (1960): 388–89.

Block, W. T. *Early Sawmill Towns of the Louisiana-Texas Borderlands.* Woodville, TX: Dogwood Press, 1996.

Borlick, J. "NC, VA Seek to Stop Giant Rodent Population Boom." *FOX 8*, March 17, 2012.

Boscareno, J. "The Rise and Fall of the Louisiana Muskrat, 1890–1960: An Environmental and Social History." Master's thesis, University of New Orleans, 2009.

Bounds, D. L. "Marsh Restoration: Nutria Control in Maryland." Maryland Cooperative Fish and Wildlife Research Unit, University of Maryland Eastern Shore, Princess Anne, MD (1998): 1–30.

———. "Nutria: An Invasive Species of National Concern." *Wetland Journal* 12 (2000): 9–16.

Bounds, D. L., M. H. Sherfy, and T. A. Mollett. "Nutria." In *Wild Mammals of North America: Biology, Management, and Conservation*, edited by G. A. Feldhamer, B. C. Thompson, and J. A. Chapman, 1119–47. Baltimore: Johns Hopkins University Press, 2003.

Bourdelle, E. "American Mammals Introduced into France in the Contemporary Period, Especially *Myocastor* and *Ondatra*." *Journal of Mammalogy* 20 (1939): 287–91.

Boyd, C. E. "Chemical Analyses of Some Vascular Aquatic Plants." *Archiv fur Hydrobiologie* 67 (1970): 78–85.

Boyd, R., and S. Penland. "A Geomorphologic Model for Mississippi River Delta Evolution." *Transactions—Gulf Coast Association of Geological Societies* 38 (1988): 443–52.

Bradshaw, J. "Flickering Candles Illuminated Mine When Journalist Visited." *Advertiser* (Lafayette, LA), April 21, 2004.

Branch, E. D. *The Hunting of the Buffalo.* 1929. Repr., Lincoln: University of Nebraska Press, 1962.

Brande, W. T. *A Dictionary of Science, Literature, and Art: Comprising the History, Description, and Scientific Principles of Every Branch of Human Knowledge; with the Derivation and Definition of All the Terms in General Use*, 2nd ed. London: Longman, Brown, Green, and Longmans, 1852.

Branan, W. "Muskrat vs. Mink." *Louisiana Conservation Review* (Winter 1937–38): 61–65.

———. "Storm over Muskrat Land." *Louisiana Conservation Review* (1940): 11.

Brewer, J. S., and J. B. Grace. "Plant Community Structure in an Oligohaline Tidal Marsh." *Vegetation* 90 (1990): 93–107.

Brochet, J. *La Chasse aux Canards: Tonne, Hutte, Cabane et Gabion du Grand Gibier Blesse.* Paris: Editions de Gerfaut, 2002.

Broekhuizen, S. *De Beverrat (Myocastor coypus) in Nederland.* Amsterdam: Rijksinstituut voor Natuurbeheer, 1977.

Brote, H. C. "Cargo Loaded Northbound." SS *Del Norte*, voyage 9, June 8–August 30, 1933, Henry Conrad Brote Collection, acc. no. 195, Earl K. Long Library, University of New Orleans, New Orleans, LA.

———. "From: Lieut. Comdr. Henry Conrad Brote, D-M, USNR." SS *Del Norte*, voyage 12, February 12–May 9, 1934, Henry Conrad Brote Collection, acc. no. 195, Earl K. Long Library, University of New Orleans, New Orleans, LA.

Brown, D. *Bury My Heart at Wounded Knee: An Indian History of the American West.* New York: Holt, Rinehart and Winston, 1971.

Brown, L. E. "An Electrophoretic Comparison of the Serum Proteins of Fetal and Adult Nutria (*Myocastor coypus*)." *Comparative Biochemistry and Physiology* 19A (1966): 479–81.

Brown, L. N. "Ecological Relationships and Breeding Biology of the Nutria (*Myocastor coypus*) in the Tampa, Florida, Area." *Journal of Mammalogy* 56 (1975): 928–30.

Buffinton, A. H. "New England and the Western Fur Trade." In *Publications of the Colonial Society of Massachusetts.* Vol. 18, *Transactions 1915–1916.* Boston: Colonial Society of Massachusetts, 1917.

Burke, J. W., and R. C. Junge. "A New Type of Water Dermatitis in Louisiana." *Southern Medicine Journal* 53 (1960): 716–19.

Burns, A. C. "Frank B. Williams: Cypress Lumber King." *Journal of Forest History* (July 1980): 127–33.

Buskey, N. "Can Nutria Fur Make a Fashionable Comeback?" *Houma (LA) Courier*, September 21, 2009.

———. "In the Market for Nutria? You'll Have to Hunt Your Own." *Daily Comet* (Houma, LA), November 18, 2008.

Cabrera, A., and J. Yepes. *Mamiferos Sud-Americanos (Vida, Costumbres y Descripcion).* Buenos Aires: Compania Argentina de Editores, 1940.

Cain, M. L., W. D. Bowman, and S. D. Hacker. *Ecology.* New York: Sinauer, 2011.

Callahan, C. R., A. P. Henderson, M. S. Eackles, and T. L. King. "Microsatellite DNA Markers for the Study of Population Structure and Dynamics in Nutria (*Myocastor coypus*)." *Molecular Ecology Notes* 5 (2005): 124–26.

Campbell, H. A., and K. P. Link. "Studies on the Hemorrhagic Sweet Clover Disease: The Isolation and Crystallization of the Hemorrhagic Agent." *Journal of Biological Chemistry* 138 (1941): 21–33.

Campbell, M. W. *The North West Company.* Toronto: Macmillan Company of Canada, 1957.

Canover, G., R. Simmonds, and M. Whalen, eds. *Management and Control Plan for Bighead, Black, Grass, and Silver Carps in the United States.* Washington, DC: Asian Carp Working Group, Aquatic Nuisance Species Task Force, 2007.

Capitol News Bureau. "House Votes to Back Wetlands Cypress Logging." *Baton Rouge Advocate*, n.d.

Carr, M. "Nutria Tales: The Rat's out of the Bag—Tabasco Mogul Didn't Bring Rodents Here." *New Orleans Times-Picayune*, September 29, 2002.

Carroll, L. *Alice's Adventures in Wonderland.* Clinton, MA: Colonel Press, 1920.

———. *Through the Looking Glass and What Alice Found There.* London: MacMillan, 1871.

Carter, J., and B. P. Leonard. "A Review of the Literature on the Worldwide Distribution, Spread of, and Efforts to Eradicate the Coypu (*Myocastor coypus*)." *Wildlife Society Bulletin* 30 (2002): 162–75.

Cave, A. A. *The French and Indian War.* Westport, CT: Greenwood Press, 2004.

Chabreck, R. H. "Daily Activity of Nutria in Louisiana." *Journal of Mammalogy* 43 (1962): 337–44.

———. "Vegetation, Water and Soil Characteristics of the Louisiana Coastal Region." Louisiana State University, Agricultural Experiment Station Bulletin No. 664 (1972).

Chabreck R. H., G. Linscombe, and N. Kinler. "Winter Food of River Otters from Saline and Fresh Environments in Louisiana." *Proceedings of the Annual Conference, Southeast Association of Fish and Wildlife* 36 (1982): 373–83.

Chabreck, R. H., and J. R. Love. "Food Habits of Nutria." *Proceedings of the 13th Annual Conference, Society of Wetland Scientists* (1993): 76–78.

Chabreck, R. H., J. R. Love, and G. Linscombe. "Foods and Feeding Habits of Nutria in Brackish Marsh in Louisiana." In *Worldwide Furbearer Conference Proceedings*, edited by J. A. Chapman and D. Pursley, 1–543. Frostburg, MD, 1981.

Chapman, J. A., G. R. Willner, K. R. Dixon, and D. Pursley. "Differential Survival Rates among Leg-Trapped and Live-Trapped Nutria." *Journal of Wildlife Management* 42 (1978): 926–28.

Chase, R. *The Complete Book of Oscar Fashion: Variety's 15 Years of Glamour on the Red Carpet.* New York: Reed Press, 2003.

Chenrow, F., and C. Chenrow. *Reading Exercises in Black History.* Elizabethtown, PA: Continental Press, 1974.

Chittenden, H. M. *The American Fur Trade of the Far West.* 2 vols. Repr., Stanford, CA: Academic Reprints, 1954.

———. *The American Fur Trade of the Far West*, Bison ed. Lincoln: University of Nebraska Press, 1986.

———. *History of Early Steamboat Navigation on the Missouri River: Life and Adventures of Joseph LaBarge.* 2 vols. New York: Francis P. Harper, 1903.

Clark, F. *Hats.* London: Anchor Press, 1992.

Clay, R. T., and W. R. Clark. "Demography of Muskrats on the Upper Mississippi River." *Journal of Wildlife Management* 49 (1985): 883–90.

Cocchi, R., and F. Riga. "Nutria *Myocastor coypus* (Molina, 1782)." *Iconografia dei Moammiferi d'Italia Isituto Nazionale per la Fauna Selvatica* 38 (1999): 1–139.

Cole, J. *Best-Loved Folktales of the World*. New York: Anchor Books, 1982.

Coleman, J. M., H. H. Roberts, and G. W. Stone. "Mississippi River Delta: An Overview." *Journal of Coastal Restoration* 14 (1998): 698–716.

Colton, C., ed. *Transforming New Orleans and Its Environs: Centuries of Change*. Pittsburgh, PA: University of Pittsburgh Press, 2000.

———. *An Unnatural Metropolis: Wrestling New Orleans from Nature*. Baton Rouge: Louisiana State University Press, 2005.

"Commercial Value of 1928–29 Fur Crop Greatest Ever Harvested." *Louisiana Conservation News* 4 (July–August 1929): 19–20.

Conaway, J. "On Avery Island, Tabasco Sauce Is the Spice of Life." *Smithsonian* 15 (1984): 72–76.

Conner, W. H. "Cypress Regeneration in Coastal Louisiana and the Impact of Nutria upon Seedling Survival." *Proceedings of the 13th Annual Conference, Society of Wetland Scientists* (1993): 130–36.

———. "The Nutria Problem—Part III: Reply to Rebuttal." *Aquaphte* 9 (1989): 14.

Conner, W. H., and J. W. Day Jr. "Productivity and Composition of a Bald Cypress–Water Tupelo Site and a Bottomland Hardwood Site in a Louisiana Swamp." *American Journal of Botany* 63 (1976): 1354–64.

———. "Rising Water Levels in Coastal Louisiana: Implications for Two Coastal Forested Wetland Areas in Louisiana." *Journal of Coastal Restoration* 4 (1988): 589–96.

Conner, W. H., and J. R. Toliver. "The Problem of Planting Cypress in Louisiana Swamplands When Nutria (*Myocastor coypus*) Are Present." *Proceedings of the Eastern Wildlife Damage Control Conference* 3 (1987): 42–49.

———. "Vexar Seedling Protectors Did Not Reduce Nutria Damage to Planted Baldcypress Seedlings." *Tree Planters' Notes* 38 (1987): 26–29.

Conrad, G. R., ed. *The Cajuns: Essays of Their History and Culture*. USL History Series, no. 11. Lafayette: Center for Louisiana Studies at the University of Southwestern Louisiana, 1983.

Contreras, L. C. "Bioenergetics of Huddling: Test of a Psycho-Physiological Hypothesis." *Journal of Mammalogy* 65 (1984): 256–62.

Cooper, C. "Louisiana Is Trying to Turn Pest into a Meal." *New York Times*, December 14, 1997.

Coreil, P. D., and H. R. Perry Jr. "A Collar for Attaching Radio Transmitter to Nutria." *Proceedings of the Southeastern Association of Game and Fish Commissioners* 31 (1977): 254–58.

Costanza, G. *Il Tecnico Operaio Conciatore e Pellicciaio*. Milan: Ulrico Hoepli Editore, 1982.

Cronon, W. *Changes in the Land: Indians, Colonists, and the Ecology of New England*. New York: Hill and Wang, 1983.

Daspit, A. P. "Activities of Fur and Wildlife Refuge Division." *Louisiana Conservationist* 1 (1943): 5.

———, Department of Conservation, to Edward Avery McIlhenny, October 16, 1930, E. A. McIlhenny Collection, Avery Island, LA.

———. "Development of Nutria in Few Years since Its Introduction in Louisiana Has Been Virtually Phenomenal." *Louisiana Game, Fur, and Fish* 6 (1947): 4.

———. "Fur Harvest Begins." *Louisiana Conservationist* 2 (1949): 17–19.

———. "Louisiana's Fur Industry." *Louisiana Conservationist* 1 (1948): 2–3.

———. "'Migrating' Nutria." *Louisiana Conservationist* 2, no. 5 (January 1950): 7–8.

———. "Trappers Pile Up Big Tale of Woe." *Louisiana Conservationist* 1, no. 7 (March 1949): 13–16.

Davis, D. W. "Historical Perspective on Crevasses, Levees, and the Mississippi River." In *Transforming New Orleans and Its Environs: Centuries of Change*, edited by C. E. Colten, 84–108. Pittsburgh, PA: University of Pittsburgh Press, 2000.

Day, J. H. *Estuarine Ecology*. Rotterdam: A. A. Balkema, 1981.

de Azara, Félix. *The Natural History of the Quadrupeds of Paraguay and the River La Plata*. Paris: Walckenaer, 1809.

Deems, E. F., and D. Pursley, eds. *North American Furbearers: Their Management, Research, and Harvest Status in 1976*. College Park: University of Maryland Press, 1978.

Defenders of Wildlife. *Changing U.S. Trapping Policy: A Handbook for Activists*. Washington, DC: Defenders of Wildlife, 1984.

deSoriano, B. S. "Elementos Constitutivos de una Habitación de *Myocastor coypus bonariensis* (Geoffrey) ('Nutria')." *Revista de la Facultad de Humanidadas y Ciencias serie Ciencias Biológicas* 18 (1960): 257–76.

Deutsch, H. B. "The Bird That's Not on Nellie's Hat." *Saturday Evening Post*, October 14, 1939, 1–3.

Dickson, H. "The Promotion of Bird City." *Saturday Evening Post*, February 10, 1984, 1–3.

"Dignitaries Give Louisiana Muskrat Approval at Washington Luncheon." *Louisiana Conservationist* 1, no. 3 (February 1943): 1–7.

Dixon, K. R., G. R. Willner, J. A. Chapmen, W. C. Lane, and D. Pursley. "Effects of Trapping and Weather on Body Weights of Feral Nutria in Maryland." *Journal of Applied Ecology* 16 (1979): 69–76.

Dolin, E. J. *Fur, Fortune, and Empire*. New York: W. W. Norton, 2010.

Doncaster, C. P., and T. Micol. "Annual Cycle of a Coypu (*Myocastor coypus*) Population: Male and Female Strategies." *Journal of Zoology (London)* 217 (1989): 227–40.

———. "Response by Coypus to Catastrophic Events of Cold and Flooding." *Holarctic Ecology* 13 (1990): 98–104.

Donlan, J., and J. Donlan. "Trappers of the Barataria: Nutria Traditions." Louisiana Department of the Arts, Office of Cultural Affairs, http://www.louisianafolklife.org/LT/Articles_Essays/trappers1.html.

Dozier, H. L. "The Present Status and Future of Nutria in the Southeast States." *Proceedings of the Annual Conference of the Southeastern Association of Game and Fish Commissioners* 5 (1985): 368–73.

———. "Suggested Muskrat Recipes." *Louisiana Conservationist* 2, no. 2 (January 1944): 1–4.

Dranguet, C. A., Jr. and R. J. Heleniak. "Man in the Basin: Habituation and Forest Exploitation in the Pontchartrain Basin." *Basics of the Basin Research Symposium Addressing the Condition of the Lake Pontchartrain Basin*, University of New Orleans, May 14, 1992.

Drew, M. C. "Soil Aeration and Plant Root Metabolism." *Soil Science* 154 (1992): 259–68.

Dunlap, T. R. "Remaking the Land: The Acclimatization Movement and Anglo Ideas of Nature." *Journal of World History* 8 (1997): 303–19.

Dyhouse, C. "Skin Deep: The Fall of Fur." *History Today* 61, no. 11 (2011): 26–29.

Ehrlich, S. "The Biology of the Nutria." *Bamidgeh* 10 (1958): 36–43, 60–70.

———. "Experiment on the Adaptation of Nutria to Winter Conditions." *Journal of Mammalogy* 43 (1962): 418.

———. "Field Studies in the Adaptation of Nutria to Seasonal Variations." *Mammalia* 31 (1967): 347–60.

Ehrlich, S., and K. Jedynak. "Nutria Influence on a Bog Lake in Northern Pomorze, Poland." *Hydrogiologia* 19 (1962): 273–97.

Ellis, E. A. *The Broads.* London: Collins, 1965.

———. "Some Effects of Selective Feeding by the Coypu (*Myocastor coypus*) on the Vegetation of Broadland." *Transactions of the Norfolk and Norwich Naturalists' Society* 20 (1963): 32–35.

Elsey, R. M. "Changes in Louisiana's Alligator Management Program." *Proceedings of the Southeastern Association of Game and Fish Commissioners* 9 (2000): 185–88.

Elsey, R. M., L. McNease, T. Joanen, and N. Kinler. "Food Habits of Native and Farm-Released Juvenile Alligators." *Proceedings of the Annual Conference of the Southeastern Association of Fish and Wildlife Agencies* 46 (1992): 57–66.

Emanuel, K. "Increasing Destructiveness of Tropical Cyclones over the Past 30 Years." *Nature* 436 (2005): 686–88.

Emberly, J. V. *The Cultural Politics of Fur.* Montreal: McGill–Queen's University Press, 1997.

Emery, T. "Killed by the Thousands, Varmint Will Never Quit." *New York Times*, July 5, 2012.

Endangered Species Act of 1973 (ESA). 16 U.S.C.A. §§ 1531–44.

Ensminger, A. B. "The Economic Status of Nutria in Louisiana." Louisiana Wildlife and Fisheries Commission, New Orleans (1955): 185–88.

———. *The Fur Animals, the Alligator, and the Fur Industry in Louisiana.* Fur and Refuge Division, Louisiana Department of Wildlife and Fisheries. Baton Rouge: Moran Colorgraphic, 1982.

Ericsson, R. J. "Male Antifertility Compounds: U-5897 as a Rat Chemosterilant." *Journal of Reproduction and Fertility* 2 (1970): 213–22.

Ericsson, R. J., H. E. Downing, R. E. Marsh, and W. E. Howard. "Bait Acceptance by Rats of Microencapsulated Male Sterilant Alpha-Chlorohydrin." *Journal of Wildlife Management* 35 (1971): 573–76.

Errington, P. L. "Drowning as a Cause of Mortality in Muskrats." *Journal of Mammalogy* 18 (1937): 497–500.

Estlack, R. W. *The Aleut Internments of World War II: Islanders Removed from Their Homes by Japan and the United States.* Jefferson, NC: MacFarland, 2014.

Evans, J. "About Nutria and Their Control." *United States Bureau of Sport Fisheries and Wildlife, Resource Publication* 86 (1970): 1–65.

———. "Nutria." In *Prevention and Control of Wildlife Damage*, edited by R. M. Timm, B61–B70. Lincoln: Cooperative Extension Service, University of Nebraska, 1983.

Evans, J., J. O. Ellis, R. D. Nass, and A. L. Ward. "Techniques for Capturing, Handling, and Marking Nutria." *Proceedings of the Southeastern Association of Game and Fish Commissioners* 25 (1971): 295–315.

Evans, J., and A. L. Ward. "Secondary Poisoning Associated with Anticoagulant-Killed Nutria." *Journal of the American Veterinary Medical Association* 151 (1967): 856–61.

Evers, D. E., J. G. Gosselink, C. E. Sasser, and J. M. Hill. "Wetland Loss Dynamics in Southwestern Barataria Basin, Louisiana (USA), 1945–1985." *Wetlands Ecology and Management* 2 (1992): 103–18.

Evers, D. E., G. O. Holm Jr., and C. E. Sasser. "Digitization of the Floating Marsh Maps in the Barataria and Terrebonne Basins, Louisiana." BTNEP Publ. No. 28 (1996), Barataria-Terrebonne National Estuarine Program, Thibodaux, LA.

Evers, D. E., C. E. Sasser, J. G. Gosselink, D. A. Fuller, and J. M. Visser. "The Impact of Vertebrate Herbivores on Wetland Vegetation in Atchafalaya Bay, Louisiana." *Estuaries* 21 (1998): 1–13.

Farber, S. "Welfare Loss of Wetlands Disintegration: A Louisiana Study." *Contemporary Economic Policy* 14 (1996): 92–106.

Farber, S., and R. Costanza. "The Economic Value of Wetland Systems." *Journal of Environmental Management* 24 (1987): 41–51.

"Fats from Fur Animals Valuable." *Louisiana Conservationist* 1, no. 2 (January 1943): 1–5.

Fears, D. "In Maryland, a Renewed Effort to Eradicate Swamp Rats from the Delmarva Peninsula." *Washington Post*, October 29, 2011.

Federal Leadership Committee for the Chesapeake Bay. "Strategy for Protecting and Restoring the Chesapeake Bay Watershed." Executive Order 13508, May 12, 2010.

Federspiel, M. N. "Nutria Farming." *American Fur Breeder* 13 (1941): 12–13.

Feldhammer, G. A., B. C. Thompson, and J. A. Chapman, eds. *Wild Mammals of North America: Biology, Management, and Conservation*. Baltimore: Johns Hopkins University Press, 2003.

Felsher, J. N. "Last Days of the Trapper." *Louisiana Life* 20 (2000): 54–57.

Ferrante, F. L. "Oxygen Conservation during Submergence Apnea in a Diving Mammal, the Nutria." *American Journal of Physiology* 218 (1970): 363–71.

Fichet-Calvet, E. "Persistence of a Systematic Labeling in Fur and Guard Hairs by Ingestion of Rhodamine B in *Myocastor coypus* (Rodentia)." *Mammalia* 63 (1999): 241–44.

Finckbeiner, S. M. "Partial Characterization of Scent Gland Compounds and a New Technique for Computer-Aided Photographic Identification of Individual Coypu." Master's thesis, Cornell University, 2005.

Fisher, R. H. *The Russian Fur Trade, 1550–1700*. Berkeley: University of California Press, 1943.

Folkow, B., B. Lisander, and B. Oberg. "Aspects of the Cardiovascular Nervous Control in a Mammalian Diver (*Myocastor coypus*). *Acta Physiologica Scandinavica* 82 (1971): 439–46.

Foote, A. L., and L. A. Johnson. "Plant Stand Development in Louisiana Coastal Wetlands: Nutria Grazing Effects on the Plant Biomass." *Proceedings of the 13th Annual Conference, Society of Wetland Scientists* (1993): 265–71.

Ford, M. A., and J. B. Grace. "Effects of Vertebrate Herbivores on Soil Processes, Plant Biomass, Litter Accumulation and Soil Elevation Changes in a Coastal Marsh." *Journal of Ecology* 86 (1998): 974–82.

———. "The Interactive Effects of Fire and Herbivory on a Coastal Marsh in Louisiana." *Wetlands* 18 (1998): 1–8.

Freeman, H. "Fighting the Return of Fur." *Guardian*, September 18, 2006.

Fricker, D. *Historic Context: The Louisiana Lumber Boom, c. 1880–1925*. Baton Rouge, LA: Historic Preservation Services.

Fuller, D. A, C. E. Sasser, W. B. Johnson, and J. G. Gosselink. "The Effects of Herbivory on Vegetation on Islands in Atchafalaya Bay, Louisiana." *Wetlands* 4 (1985): 105–14.

Gabrey, S. W., N. Kinler, and R. M. Elsey. "Impacts of Nutria Removal on Food Habits of American Alligators in Louisiana." *Southeastern Naturalist* 8 (2009): 347–54.

Galluci, M. "BP Oil Spill: Louisiana Wetland Loss Is Speeding Up Due to Crude from Deepwater Horizon Disaster." *International Business Times*, April 15, 2015.

Garland, H. *Main-Travelled Roads*. New York: Macmillan, 1981.

Garner, R. *The Political Theory of Animal Rights*. Manchester, UK: Manchester University Press, 2005.

Garrett, M. G., and W. L. Franklin. "Diethylstilbestrol as a Temporary Chemosterilant to Control Black-Tailed Prairie Dog Populations." *Journal of Range Management* 36 (1983): 753–56.

Gayarré, C. E. *History of Louisiana*. Vol. 2. Gretna, LA: Pelican, 1974.

Gebhardt, H. "Ecological and Economic Consequences of Introductions of Exotic Wildlife (Birds and Mammals) in Germany." *Wildlife Biology* 2 (1996): 205–11.

Geoffroy Saint-Hilaire, É. "Mémoire sur un Nouveau Genre de Mammifères Nommé *Hydromys*." *Annales de Musée National d'Historie Naturelle, Paris* 6 (1805): 81–90.

George, W., and B. J. Weir. "Hystricomorph Chromosomes." *Symposia of the Zoological Society of London* 34 (1974): 79–108.

Gersberg, R. M., B. V. Elkins, S. Lyon, and C. R. Goldman. "Role of Aquatic Plants in Wastewater Treatment by Artificial Wetlands." *Water Research* 20 (March 1986): 363–68.

Gilbert, B. *Westering Man: The Life of Joseph Walker*. Norman: University of Oklahoma Press, 1985.

Giles, L. W., and V. L. Childs. "Alligator Management of the Sabine National Wildlife Refuge." *Journal of Wildlife Management* 13 (1949): 16–28.

Gill, J. A. "The Buffer Effect and Large-Scale Population Regulation in Migratory Birds." *Nature* 412 (2001): 436–38.

Gill, R. R. "Wildlife Research: An Endangered Species." *Wildlife Society Bulletin* 13 (1985): 580–87.

Gomez, G. M. "Perspective, Power, and Priorities: New Orleans and the Mississippi River Flood of 1927." In *Transforming New Orleans Environs: Centuries of Change*, edited by C. Colton, 117–20. Pittsburgh, PA: University of Pittsburgh Press, 2000.

———. *A Wetland Biography: Seasons on Louisiana's Chenier Plain*. Austin: University of Texas Press, 1998.

Gosling, L. M. "Climatic Determinants of Spring Littering by Feral Coypus, *Myocastor coypus*." *Journal of Zoology (London)* 195 (1981): 281–88.

———. "The Coypu in East Anglia." *Transactions of the Norfolk and Norwich Naturalists' Society* 23 (1974): 49–59.

———. "Coypu." In *The Handbook of British Mammals*, 2nd ed., edited by G. B. Corbet and H. N. Southern, 256–65. Oxford: Blackwell Scientific Press, 1977.

———. "The Duration of Lactation in Feral Coypus (*Myocastor coypus*)." *Journal of Zoology (London)* 191 (1980): 461–74.

———. "The Dynamics and Control of a Feral Coypu Population." In *Proceedings of the Worldwide Furbearer Conference*, edited by J. A. Chapman and D. Pursley, 1806–25. Frostburg, MD: Worldwide Furbearer Conference, 1981.

———. "The Effect of Cold Weather on Success in Trapping Feral Coypus (*Myocastor coypus*)." *Journal of Applied Ecology* 18 (1981): 467–70.

———. "Extinction to Order." *New Scientist (London)* 121 (1989): 44–49.

———. "Selective Abortion of Entire Litters in the Coypu (*Myocastor coypus*): Adaptive Control of Offspring Production in Relation to Quality and Sex." *American Naturalist* 127 (1986): 772–95.

———. "Towards an Eradication Plan for Nutria in Maryland." *Report to the Maryland Department of Natural Resources* (2001): 1–14.

———. "The Twenty-Four Hour Activity Cycle of Captive Coypus *Myocastor coypus*." *Journal of Zoology (London)* 187 (1979): 341–67.

Gosling, L. M., and S. J. Baker. "Coypu (*Myocastor coypus*) Potential Longevity." *Journal of Zoology (London)* 197 (1981): 285–312.

———. "The Eradication of Muskrats and Coypus from Britain." *Biological Journal of the Linnean Society* 38 (1989): 39–51.

———. "Planning and Monitoring an Attempt to Eradicate Coypus from Britain." *Symposium of the Zoological Society of London* 58 (1987): 99–113.

Gosling, L. M., S. J. Baker, and K. M. H. Wright. "Differential Investment by Female Coypus (*Myocastor coypus*) during Lactation." *Symposia of the Zoological Society of London* 51 (1984): 273–300.

Gosling, L. M., A. D. Watt, and S. J. Baker. "Continuous Retrospective Census of the East Anglian Coypu Population between 1970–1979." *Journal of Animal Ecology* 50 (1981): 885–901.

Gosling, L. M., K. M. H. Wright, and G. D. Few. "Facultative Variation in the Timing of Parturition by Female Coypus (*Myocastor coypus*), and the Cost of Delay." *Journal of Zoology (London)* 214 (1988): 407–15.

Gosselink, J. G., J. M. Coleman, and R. E. Stewart Jr. "Coastal Louisiana." In *Status and Trends of the Nation's Biological Resources*, edited by M. J. Mac, P. A. Opler, C. E. Puckett Haecker, and P. D. Doran, 385–436. Reston, VA: US Department of the Interior, US Geological Survey, 1998.

Gough, L., and J. B. Grace. "Effects of Flooding, Salinity, and Herbivory on Coastal Plant Communities, Louisiana, United States." *Oecologia* 177 (1998): 527–35.

———. "Herbivore Effects on Plant Species Density at Varying Productivity Levels." *Ecology* 79 (1998): 1586–94.

Gowanloch, J. N. "Have You Ever Eaten Muskrat?" *Louisiana Conservation Review* 7 (1938): 130.

———. "Louisiana's Muskrat Industry as a Source of Human Food." *Louisiana Conservationist* 1 (1943): 4–8.

Grace, J. B. "The Impact of Nutria (*Myocastor coypus*) on Gulf Coastal Wetlands Symposium Introduction." *Proceedings of the 13th Annual Conference, Society of Wetland Scientist*s (1993): 70–74.

Grace, J. B., and M. A. Ford. "The Potential Impact of Herbivory on the Susceptibility of the Marsh Plant *Sagittaria lancifolia* to Saltwater Intrusion in Coastal Wetlands." *Estuaries* 19 (1996): 13–20.

Grace, J. B., B. D. Marx, and K. L. Taylor. "The Effects of Herbivory on Neighbor Interactions along a Coastal Marsh Gradient." *American Journal of Botany* 84 (1997): 709–15.

Grace, R. M. "Is Tabasco Sauce Patterned After Col. White's 'Tobasco Extract'?" *Metropolitan News Enterprise* (Los Angeles), July 22, 2004.

———. "More Tabasco Sauce Lore Appears to Be Fiction." *Metropolitan News Enterprise* (Los Angeles), July 29, 2004.

———. "Tabasco: A Hot Sauce with an Uncertain Background." *Metropolitan News Enterprise* (Los Angeles), July 8, 2004.

———. "Was Col. Maunsel White the True Originator of Tabasco Sauce?" *Metropolitan News Enterprise* (Los Angeles), July 15, 2004.

Grant, H. "Revenge of the Paris Hat: The European Craze for Wearing Headgear Had a Profound Effect on Canadian History." *Beaver* (1989): 3744.

Grant, W. B., to J. Dymond, president of Delta Duck Club, October 16, 1926, E. A. McIlhenny Collection, Avery Island, LA.

Greer, J. K. "Mammals of Malleco Province Chile." *Publications of the Museum, Michigan State University Biological Series* 3 (1966): 49–152.

Griffis, W. E. *The Story of New Netherland*. New York: Houghton Mifflin, 1909.

Griffo, J. V. "The Status of Nutria in Florida." *Journal of the Florida Academy of Science* 20 (1957): 209–15.

Grimm (The Brothers Grimm). *Deutsche Sagen*. Berlin: Mittel Strasse no. 6, 1816.

Guichon, M. L., and M. H. Cassini. "Local Determinants of Coypu Distribution along the Lujan River, East Central Argentina." *Journal of Wildlife Management* 63 (1999): 895–900.

Hall, E. R. *The Mammals of North America*, 2nd ed. New York: John Wiley and Sons, 1981.

Hammond, E. E., L. E. Nolfo, J. Carter, S. Merino, and R. F. Aguilar. "Surgical Implantation of Abdominal Radio Transmitters in Wild Nutria (*Myocastor coypus*) in Southern Louisiana." Society of Wetland Scientist Annual Meeting, New Orleans, LA, June 7–13, 2013.

Hammond, L. "Market Wildlife: The Hudson's Bay Company and the Pacific Northwest, 1821–49. *Forest and Conservation History* 37 (1993): 14–25.

Hansen, H. H. *Costume Cavalcade: 685 Examples of Historic Costume in Colour*. London: Methuen, 1956.

Han, S. H., D. H. Kong, and J. Y. Cha. *Exotic Animal Species in South Korea*. Ecological Society of America 84th Annual Meeting, Spokane, WA, August 8–12, 1999.

Haramis, G. M., and T. S. White. "A Beaded Collar for Dual Micro GPS/VHS Transmitter Attachment to Nutria." *Mammalia* 75 (2011): 79–82.

Harper, D. M., K. M. Mavuti, and S. M. Muchiri. "Ecology and Management of Lake Naivasha, Kenya, in Relation to Climatic Change, Alien Species' Introduction, and Agricultural Development." *Environmental Conservation* 17 (1990): 328–36.

Harris, B. *The Lives of Mountain Men*. Guilford, CT: Lyons Press, 2005.

Harris, S. "Nutria—An Invasive Species." *Eugene (OR) Daily News*, July 3, 2015.

Harris, V. T. "The Nutria as a Wild Fur Mammal in Louisiana." *Twenty-First North American Wildlife Conference* (1956): 474–86.

Harris, V. T., and R. H. Chabreck. "Some Effects of Hurricane Audrey on the Marsh at Marsh Island, Louisiana." *Louisiana Academy of Sciences* 21 (1958): 47–51.

Harris, V. T., and F. Webert. "Nutria Feeding Activity and Its Effect on Marsh Vegetation in Southwestern Louisiana." *United States Fish and Wildlife Service, Special Scientific Report* 64 (1962): 1–53.

Hatcher, R. T., and J. H. Shaw. "A Comparison of Three Indices to Furbearer Populations." *Wildlife Society Bulletin* 9 (1981): 153–56.

Hazardous Substances and New Organisms Act of 2003, New Zealand, http://legislation. govt.nz/act/public/1996/0030/latest/DLM386556.html#DLM386556.

Hebert, M. "Agricultural Damage." *Proceedings of the Nutria and Muskrat Management Symposium, Louisiana Cooperative Extension Service* (1992): 1–100.

Heleniak, R. J., and C. A. Dranguet. "Changing Patterns of Human Activity in the Western Basin of Lake Pontchartrain." In *Proceedings of the Tenth National Conference of the Coastal Society*, October 12–15, 1986, New Orleans.

Henderson, W. C., acting chief of the Bureau of Biological Survey, to W. B. Grant, November 17, 1926, E. A. McIlhenny Collection, Avery Island, LA.

Hillbricht, A., and L. Ryszkowski. "Investigations of the Utilization and Destruction of Its Habitat by a Population of Copyu, *Myocastor coypu* Molina, Bred in Semi-Captivity." *Ekologia Polska, Seria A* 9 (1961): 506–24.

Hilton, H. W., and W. H. Robinson. "Fate of Zinc Phosphide and Phosphine in the Soil-Water Environment." *Journal of Agriculture Food Chemistry* 20 (1972): 1209–12.

Holmes, R. G., O. Illman, and J. K. A. Beverley. "Toxoplasmosis in Coypu." *Veterinary Record* 101 (1977): 74–75.

Holm, G. O., Jr., E. Evers, and C. E. Sasser. "The Nutria in Louisiana: A Current and Historical Perspective." *Lake Pontchartrain Basin Foundation* (2011): 1–58.

Holm, G. O., Jr., C. E. Sasser, G. W. Peterson, and E. M. Swenson. "Vertical Movement and Substrate Characteristics of Oligohaline Marshes Near a High-Sediment, Riverine System." *Journal of Coastal Research* 16 (2000): 164–71.

Honeycutt, R. L. "Rodents (Rodentia)." In *The Timetree of Life*, edited by S. B. Hedges and S. Kumar, 490–94. London: Oxford University Press, 2009.

Horton, T. "Chesapeake's Marshes May See End of Invasive Nutria." Bay Journal News Service, *Southern Maryland Online*, May 20, 2014.

Housse, P. R. *Animales Salvajes de Chile en Su Clasificación Moderna: Su Vida y Costumbres*. Santiago: Ediciónes de la Universidad de Chile, 1953.

Howerth, E. W., A. J. Reeves, M. R. McElveen, and F. W. Austin. "Survey for Selected Diseases in Nutria (*Myocastor coypus*) in Louisiana." *Journal of Wildlife Diseases* 30 (1994): 450–53.

Hunter, D. *Half Moon: Henry Hudson and the Voyage That Redrew the Map of the New World*. New York: Bloomsbury Press, 2009.

Hunt, W. J., Jr. "Fort Flood: An Enigmatic Nineteenth-Century Trading Post." *North Dakota History* 61 (1994): 7–20.

Hurtado, A. L. *John Sutter: A Life on the American Frontier.* Norman: University of Oklahoma Press, 2006.

Innis, H. *The Fur Trade in Canada: An Introduction to Canadian Economic History.* Toronto: University of Toronto Press, 1970.

International Association of Fish and Wildlife Agencies (IAFWA). *Improving Animal Welfare in U.S. Trapping Programs: Process Recommendations and Summaries of Existing Data.* Washington, DC: IAFWA Fur Resources Technical Subcommittee and Trapping Work Group, 1997.

Irving, W. *The Adventures of Captain Bonneville; or, Scenes beyond the Rocky Mountains of the Far West.* 2 vols. London: Richard Bentley, 1837.

———. *Astoria; or, Enterprise beyond the Rocky Mountains.* Paris: Baudry's European Library, 1836.

Israel, B. "Swamp Rats on the Move as Winters Warm." *Scientific American,* August 12, 2013, http://www.scientificamerican.com/article/swamp-rats-on-the-move-as-winters-warm/.

Jackson, D. D. "Orangetooth Is Here to Stay." *Audubon,* July 1990, 90–94.

Jackson, W. G. "Management of Canine Mismating with Diethylstilbestrol." *California Veterinarian* 22 (1953): 29.

Jelinek, P., L. Valicek, B. Smid, and R. Halouzka. "Determination of Papillomatosis in the Coypus (*Myocastor coypus* Molina)." *Veterinarni Medicina (Prague)* 23 (1978): 113–19.

Jenkins, S. H., and P. E. Busher. "*Castor canadensis.*" *American Society of Mammalogists, Mammalian Species* 120 (1979): 1–8.

Jing, J. "Forget Phil! T-Boy the Nutria Makes His Weather Prediction." *WGNO News,* February 2, 2015, http://wgno.com/2015/02/02/forget-about-phil-t-boy-the-nutria-makes-his-weather-prediction/.

Joanen, T., and L. McNease. "The Management of Alligators in Louisiana, USA." In *Wildlife Management: Crocodiles and Alligators,* edited by G. J. W. Webb, S. C. Manolis, and P. J. Witehead, 33–42. Chipping Norton, Australia: Surrey Teatty and Sons, 1987.

Joanen, T., L. McNease, G. Perry, D. Richard, and D. Taylor. "Louisiana's Alligator Management Program." *Proceedings of the Annual Conference of the Southeastern Association of Fish and Wildlife Agencies* 38 (1984): 210–11.

Johannsen, A. *The House of Beadle and Adams and Its Dime and Nickel Novels: The Story of a Vanished Literature II.* Norman: University of Oklahoma Press, 1950.

Johnson, A. *The Swedish Settlements on the Delaware: Their History and Relation to the Indians, Dutch and English, 1638–1664.* 2 vols. New York: D. Appleton, 1999.

Johnson, S. *A Dictionary of the English Language: An Anthology.* New York: Penguin, 2005.

Jordan, J., and E. Mouton. *Coastwide Nutria Control Program 2010.* CWPPRA Project (LA-03b) Final Report, Louisiana Department of Wildlife and Fisheries, 2010.

Kanwanich, S. "The Arguments for and against Breeding Nutrias." *Bangkok Post Perspective,* February 8, 1998.

Katomski, P. A., and F. L. Ferrante. "Catecholamine Content and Histology of the Adrenal Glands of the Nutria (*Myocastor coypus*)." *Comparative Biochemistry and Physiology* 48A (1974): 539–46.

Keddy, P.A. *Water, Earth, Fire: Louisiana's Natural Heritage*. Philadelphia: Xlibris, 2008.

———. *Wetland Ecology: Principles and Conservation*, 2nd ed. Cambridge: Cambridge University Press, 2010.

Keddy, P. A., D. Campbell, T. McFalls, G. P. Shaffer, R. Moreau, C. Dranguet, and R. Heleniak. "The Wetlands of Lakes Pontchartrain and Maurepas: Past, Present and Future." *Environmental Reviews* 15 (2007): 43–77.

Keddy, P. A., L. Gough, J. A. Nyman, T. McFalls, J. Carter, and J. Siegrist. "Alligator Hunters, Pelt Traders, and Runaway Consumption of Gulf Coast Marshes." In *Human Impacts on Salt Marshes: A Global Perspective*, edited by B. R. Silliman, Mark D. Bertness, and Edwin D. Grosholz, 115–33. Berkeley: University of California at Berkeley Press, 2009.

Kendrot, S. "Eradication Strategies for Nutria in Delaware and Chesapeake Bay Wetlands: Annual Report, September 1, 2002–August 31, 2003." Annapolis, MD: US Department of Agriculture, 2004.

Kennedy, M. L., and P. K. Kennedy. "First Record of Nutria, *Myocastor coypus* (Mammalia: Rodentia) in Tennessee." *Brimleyana* 25 (1998): 156–57.

Kerr, R. *The Animal Kingdom; or, Zoological System, of the Celebrated Sir Charles Linnaeus*. London: Murray and R. Faulder, 1792.

Kinler, N., and G. Linscombe. *A Survey of Nutria Herbivory Damage in Coastal Louisiana in 1998*. Baton Rouge: Fur and Refuge Division, Louisiana Department of Wildlife and Fisheries, 1998.

Kinler, N., R. G. Linscombe, and R. H. Chabreck. "Smooth Beggartick, Its Distribution, Control and Impact on Nutria in Coastal Louisiana." In *Worldwide Furbearer Conference Proceedings*, edited by J. A. Chapman and D. Pursley. Frostburg, MD: 1980.

Kinler, N., G. Linscombe, and S. Hartley. "A Survey of Nutria Herbivory Damage in Coastal Louisiana in 2000." Fur and Refuge Division, Louisiana Department of Wildlife and Fisheries, 2000.

Kinler, N., G. Linscombe, and P. R. Ramsey. "Nutria." In *Wild Furbearer Management and Conservation in North America*, edited by M. Novak, J. A. Baker, M. E. Obbard, and B. Malloch, 331–43. Toronto: Ontario Trappers Association, Ontario Ministry of Natural Resources, 1987.

Kinler, Q. J., R. H. Chabreck, N. W. Kinler, and R. G. Linscombe. "Effect of Tidal Flooding on Mortality of Juvenile Muskrats." *Estuaries* 13 (1990): 337–40.

Knight, C. *Sketches in Natural History: History of the Mammalia (in Six Volumes)*. London: W. Clowes and Sons, 1849.

Kruse, R. "The Impact of Nutria (*Myocastor coypus*) as an Invasive Species and Its Possible Distribution in Washington State." Master's thesis, Evergreen State College, 2012.

Kuhn, L. W., and E. P. Peloquin. "Oregon's Nutria Problem." *Vertebrate Pest Conference* 6 (1974): 101–5.

Lake Pontchartrain Basin Foundation (LPBF). *Comprehensive Management Plan for Habitats in the Lake Pontchartrain Basin*. New Orleans: LPBF, 2005.

Lamb, B. L., R. P. Reading, and W. F. Andelt. "Attitudes and Perceptions about Prairie Dogs." In *Conservation of the Black-Tailed Prairie Dog—Saving North America's Western Grasslands*, edited by J. L. Hoogland, 108–14. Washington, DC: Island Press, 2008.

Lane, R. "The Treasures of Tabasco: Collectors Swarm Estate of John S. McIhenny." *Maine Antiques Digest*, 1998.

Lane, R. R., J. W. Day, B. Marx, E. Reyes, and G. P. Kemp. "Seasonal and Spatial Water Quality Changes in the Outflow Plume of the Atchafalaya River, Louisiana, USA." *Estuaries* 25 (2002): 30–42.

Lane, R. R., H. S. Mashriqui, G. P. Kemp, J. W. Day, J. N. Day, and A. Hamilton. "Potential Nitrate Removal from a River Diversion into a Mississippi Delta Forested Wetland." *Ecological Engineering* 20 (2003): 237–49.

Langer, L. N. *Historical Dictionary of Medieval Russia*. Lanham, MD: Scarecrow Press, 2002.

Larivière, S. "*Mustela vison.*" *American Society of Mammalogists, Mammalian Species* 608 (1999): 1–9.

Larivière, S., and L. R. Walton. "*Lontra canadensis.*" *American Society of Mammalogists, Mammalian Species* 587 (1998): 1–8.

Larrison, E. J. "Feral Coypus in the Pacific Northwest." *Murrelet* 24 (1943): 3–9.

"Last Year's Fur Crop Was Record! Progress Report of Fur and Refuge Division Submitted by Director." *Louisiana Game, Fur and Fish* 4, no. 10 (September 1946): 5.

Laurie, E. M. O. "The Coypu (*Myocastor coypus*) in Great Britain." *Journal of Animal Ecology* 15 (1946): 22–34.

LaVista, J., and G. Bodin. "How Are Louisiana's Wetlands Changing?" Press release, US Geological Survey, June 2, 2011.

LaVoie, G. F., and F. Schitoskey Jr. "Nutria Damage to Sugar Cane—A 5-Year Appraisal." *Sugar Bulletin* 10 (1973): 13–14.

Lay, D. W. "Muskrat Investigations in Texas." *Journal of Wildlife Management* 9 (1945): 56–76.

Lay, D. W., and T. O'Neill. "Muskrats on the Texas Coast." *Journal of Wildlife Management* 6 (1942): 301–11.

LeBlanc, D. J. "Nutria: Prevention and Control of Wildlife Damage." University of Nebraska-Lincoln, no. 2 (1994): B71–B80.

Le Coz, E. "Hurricane Isaac Leaves Nutria Dead on Mississippi Beaches." Reuters, September 4, 2012.

Lee, H. F. "Susceptibility of Mammalian Host to Experimental Infection with *Heterobilharzia americana.*" *Journal of Parasitology* 48 (1962): 740–45.

Lee, M. *Fashion Victims: Our Love-Hate Relationship with Dressing, Shopping, and the Cost of Style*. New York: Broadway Books, 2003.

Lewis, O. *The Big Four*. New York: Alfred Knopf, 1938.

Limerick, P. H. *The Legacy of Conquest: The Unbroken Past of the American West*. New York: W. W. Norton, 1987.

Link, R. "Living with Nutria in Washington." Washington Department of Fish and Wildlife (2006): 1–7.

Linscombe, G., N. Kinler, and V. Wright. "Nutria Population Density and Vegetative Changes in Brackish Marsh in Coastal Louisiana." In *Proceedings of the Worldwide Furbearer Conference*, edited by J. A. Chapman and D. Pursley, 129–41. Frostburg, MD: Worldwide Furbearer Conference, 1981.

Linscombe G., and E. Mouton. *Coastwide Nutria Control Program 2013–2014*. CWPPRA Project (LA-03b) Final Report, Louisiana Department of Wildlife and Fisheries, 2006.

Linscombe, R. G. "Efficiency of Padded Foothold Traps for Capturing Terrestrial Furbearers." *Wildlife Society Bulletin* 16 (1988): 307–9.

———. *The Fur Animals, the Alligator, and the Fur Industry in Louisiana.* Baton Rouge: Louisiana Department of Wildlife and Fisheries, 1980.

———. *1999–2000 Annual Report: Fur and Alligator Advisory Council.* New Iberia: Louisiana Department of Wildlife and Fisheries, 2000.

Lippson, A. J., and R. L. Lippson. *Life in the Chesapeake Bay.* Baltimore: Johns Hopkins University Press, 2006.

Litjens, B. E. J. "The Coypu, *Myocastor coypus* (Molina) in the Netherlands. I. Population Development during the Period 1963–1979." *Lutra* 23 (1980): 43–53.

———. "De Beverrat *Myocastor coypus* in Nederlands Limburg en Aangrenzende Gebieden." *Lutra* 27 (1984): 208.

Little, M. D. "Dermatitis in a Human Volunteer Infected with *Strongyloides* of Nutria and Raccoon." *American Journal of Tropical Medicine and Hygiene* 14 (1965): 1007–9.

Llewellyn, D. W., and G. P. Shaffer. "Marsh Restoration in the Presence of Intense Herbivory: The Role of *Justicia lanceolata* (Chapm.) Small." *Wetlands* 13 (1993): 176–84.

Lohmeier, L. "Home Range, Movements and Population Density of Nutria on a Mississippi Pond." *Journal of the Mississippi Academy of Science* 26 (1981): 50–54.

Lopez, B. H. *Of Wolves and Men.* New York: Scribner, 1978.

Lopez, J. A. "Chronology and Analysis of Environmental Impacts within the Pontchartrain Basin of the Mississippi Delta Plain: 1718–2002." PhD diss., University of New Orleans, 2003.

"Louisiana Declares War on the Nutria: Like Maryland, State Plans to Halt Marsh Damage by Trapping." *New York Times*, May 28, 2002.

Louisiana Department of Wildlife and Fisheries (LDWF). "Louisiana's Alligator Management Program: 2013–2014 Annual Report." *Presented to the House Committee on Natural Resources and Environment and the Senate Committee on Natural Resources* (December 2013): 1–31.

———. *Monitoring Plan: Project No. LA-02.* Baton Rouge: Nutria Harvest and Wetland Restoration Demonstration Project, 1998.

———. *Nutria: Wetland Damage.* New Orleans: Author, 2008, http://www.nutria.com/site4.php.

"Louisiana Muskrat May Help Alleviate Nation's Meat Shortage: Large Quantities Shipped to Market." *Louisiana Conservationist* 1, no. 2 (January 1943): 1–5.

Lowery, G. H. *The Mammals of Louisiana and Its Adjacent Waters.* Baton Rouge: Louisiana State University Press, 1974.

Madsen, A. *John Jacob Astor: America's First Multimillionaire.* New York: John Wiley and Sons, 2001.

Mancall, P. *Fatal Journey: The Final Expedition of Henry Hudson, a Tale of Mutiny and Murder in the Arctic.* New York: Basic Books, 2009.

Manning, R. *Grassland.* New York: Viking Press, 1995.

Manno, T. G. "Tedro's Last Stand: Hi Jolly and the U.S. Camel Corps in Arizona." *Desert Leaf,* October 2013, 47–51.

Mann, T., and E. D. Wilson. "Biochemical Observations on the Male Accessory Organs of Nutria, *Myocastor coypus* (Molina)." *Journal of Endocrinology* 25 (1962): 407–8.

Manuel, J., and E. Mouton. *Coastwide Nutria Control Program 2013–2014.* CWPPRA Project (LA-03b) Final Report, Louisiana Department of Wildlife and Fisheries, 2014.

Martin, C. *Keepers of the Game: Indian-Animal Relationships and the Fur Trade.* Berkeley: University of California Press, 1978.

Martin, W. "Visceral and Osteological Anatomy of the *Coypus* (*Myopotamus Coypus,* Comm.)." *Proceedings of the Zoological Society of London* (1835): 173–82.

Marx, J., E. Mouton, and G. Linscombe. "Nutria Harvest Distribution 2003–2004 and a Survey of Nutria Herbivory Damage in Coastal Louisiana in 2004." Fur and Refuge Division, Louisiana Department of Wildlife and Fisheries (2004): 1–45.

Matouch, O., J. Donsek, and O. Ondracek. "Rabies in the Nutria." *Veterinarstvi* 28 (1978): 549.

Matthias, K. E. K. "Nutria: A Profitable Fur Discovery." *American Fur Breeder* (1941): 18–20.

Mayer, J. J. *Wild Pigs: Biology, Damage, Control Techniques and Management.* Aiken, SC: Savannah River National Laboratory, SRNL-RP, 2009, 00869.

McFalls, T. "Effects of Disturbance and Fertility upon the Vegetation of a Louisiana Coastal Marsh." Hammond: Southeastern Louisiana University, 2004.

McFalls, T., P. A. Keddy, D. Campbell, and G. Shaffer. "Hurricanes, Floods, Levees, and Nutria: Vegetation Responses to Interacting Disturbance and Fertility Regimes with Implications for Coastal Wetland Restoration." *Journal of Coastal Research* 26, no. 5 (2010): 901–11.

McGee, H. F., Jr. "The Use of Furbearers by Native North Americans after 1500." In *Wildlife Furbearer Management and Conservation in North America,* edited by M. Novak, J. A. Baker, M. E. Obbard, and B. Malloch, 13–20. Toronto: Ontario Ministry of Natural Resources, 1987.

McHugh, J. L. "Conservation Second Only to War as Vital Problem of American People: Conservation Dept. Placed on War Time Basis." *Louisiana Conservationist* 1, no. 1 (December 1942): 1–4.

McIlhenny, E. A. *The Alligator's Life History.* Boston: Christopher, 1935. Repr., Berkeley, CA: Ten Speed Press, 1987.

———. *Befo' de War Spirituals.* Boston: Christopher, 1933.

———, to S. C. Arthur, Department of Conservation, October 29, 1926, E. A. McIlhenny Collection, Avery Island, LA.

———, to A. P. Daspit, Department of Conservation, October 18, 1930, E. A. McIlhenny Collection, Avery Island, LA.

McNease, L., N. Kinley, T. Joanen, and D. Richard. "Distribution and Relative Abundance of Alligator Nests in Louisiana Coastal Marshes." In *Crocodiles: Proceedings of the 12th Working Meeting,* 108–120. Survival Commission of IUCN–The World Conservation Union, 1994.

Mendelssohn, I. A., G. L. Andersen, D. M. Baltz, R. H. Caffey, K. R. Carman, J. W. Fleeger, S. B. Joye, et al. "Oil Impacts on Coastal Wetlands: Implications for the Mississippi River Delta Ecosystem after the Deepwater Horizon Oil Spill." *BioScience* 62 (2012): 562–74.

Merino, S., J. Carter, and G. Thibodeaux. "Testing Tail-Mounted Transmitters with *Myocastor coypus* (Nutria)." *Southeastern Naturalist* 6 (2007): 159–64.

Meyer, A. "The Impacts of Nutria on Vegetation and Erosion in Oregon." Master's thesis, State University of New York at Binghamton, 2006.

Michalski, Z., and W. Scheuring. "Coccidiosis of Intestine in the Nutria." *Wiadomosci Parazytologiczne* 25 (1979): 99–104.

Middle, B. A., ed. *Flood Pulsing in Wetlands: Restoring the Natural Hydrological Balance*. New York: John Wiley, 2002.

Mills, E. A. *In Beaver World*. 1913. Repr., Lincoln: University of Nebraska Press, 1990.

Mirsky, S. "Antigravity: Call of the Reviled." *Scientific American*, June 1, 2008, https://www.scientificamerican.com/article/call-of-the-reviled/.

Misiroglu, G. *American Counterculture: An Encyclopedia of Nonconformists, Alternative Lifestyles, and Radical Ideas in U.S. History*. New York: Routledge, 2009.

Miura, S. "Dispersal of Nutria in Okayama Prefecture." *Journal of the Mammalogical Society of Japan* 6 (1976): 231–37.

Moinard, C., C. P. Doncaster, and H. Barre. "Indirect Calorimetry Measurements of Behavioral Thermoregulation in a Semi-Aquatic Social Rodent, *Myocastor coypus*." *Canadian Journal of Zoology* 70 (1992): 907–11.

Molina, G. I. *The Geographical, Natural, and Civil History of Chili, Translated from the Original Italian by an American Gentleman*. Middleton, CT: R. Alsop, 1808.

———. *Saggio sulla Storia Naturale del Chil[e]*. Bologna, Italy: Stamperia di S. Tomnaso d'Aquino, 1782.

Mountjoy, S. *Manifest Destiny: Westward Expansion*. New York: Chelsea House, 2009.

Morgan, D. L. *Jedediah Smith and the Opening of the West*. Lincoln: University of Nebraska Press, 1964.

Morgan, L. H. *The American Beaver*. Philadelphia: L. B. Lippincott, 1868.

Morgan, R. P., II, G. R. Willner, and J. A. Chapman. "Genetic Variation in Maryland Nutria, *Myocastor coypus*." In *Proceedings of the Worldwide Furbearer Conference*, edited by J. A. Chapman and D. Pursley, 30–37. Frostburg, MD: Worldwide Furbearer Conference, 1981.

Morgan, T. D., and T. O'Neil. "Stalking the Wild Fur: A Bicentennial Report." *Louisiana Conservationist* 28 (1976): 22–27 (A#242).

Mortenson, E. "Gresham Dog Dies in Conibear Trap Set Out to Catch Nutria." *Oregonian/Oregon Live*, December 1, 2011.

Mouton, E., G. Linscombe, and S. Hartley. *A Survey of Nutria Herbivory Damage in Coastal Louisiana in 2001*. Baton Rouge: Fur and Refuge Division, Louisiana Department of Wildlife and Fisheries, 2001.

Müller-Schwarze, D., and L. Sun. *The Beaver: Natural History of a Wetlands Engineer*. Ithaca, NY: Comstock, 2003.

Murua, R., O. Neumann, and I. Dropelmann. "Food Habits of *Myocastor coypus* in Chile." In *Proceedings of the Worldwide Furbearer Conference*, edited by J. A. Chapman and D. Pursley, 544–58. Frostburg, MD: Worldwide Furbearer Conference, 1981.

"Muskrat Changes Its Name from 'Hudson Seal' to 'Southern Mink.'" *Louisiana Conservation News* 4 (March–April 1929): 11.

Musser, G. G., and M. D. Carleton. "Family Muridae." In *Mammal Species of the World: A Taxonomic and Geographic Reference*, edited by D. E. Wilson and D. M. Reeder, 501–755. Washington: Smithsonian Institution Press, 1993.

Myers, R. S., G. P. Shaffer, and D. W. Llewellyn. "Baldcypress (*Taxodium distichum* [L.] Rich.) Restoration in Southeast Louisiana: The Relative Effects of Herbivory." *Wetlands* 15 (1995): 141–48.

The Nature Conservancy. *Conservation Area Plan for the Lake Pontchartrain Estuary.* Covington, LA: North Shore Field Office, 2004.

Nelson, E. W., chief of Bureau of Biological Survey, to J. Dymond, president of Delta Duck Club, December 14, 1926, E. A. McIlhenny Collection, Avery Island, LA.

Newman, P. C. "Canada's Fur Trading Empire: Three Centuries of the Hudson's Bay Company." *National Geographic Magazine* 172 (1987): 192–228.

———. *Company of Adventurers.* Vol. 1–2. New York: Penguin Books, 1985–87.

———. *Empire of the Bay: The Company of Adventurers That Seized a Continent.* New York: Penguin Books, 1998.

Newson, R. M. "Populations Dynamics of the Coypu, *Myocastor coypus* (Molina), in Eastern England." In *Energy Flows through Small Mammal Populations*, edited by K. Petrusewicz and L. Ryszkowski, 203–4. Warsaw: Polish Scientific, 1969.

———. "Reproduction in the Feral Coypu, *Myocastor coypus.*" In *Comparative Biology of Reproduction in Mammals*, edited by I. W. Rowlands, 323–34. London: Symposia of the Zoological Society of London, 1966.

———. "Reproduction in the Feral Coypu (*Myocastor coypus*)." *Journal of Reproduction and Fertility* 9 (1965): 380–81.

Newson, R. M., and R. G. Holmes. "Some Ectoparasites of the Coypu (*Myocastor coypus*) in Eastern England." *Journal of Animal Ecology* 37 (1968): 471–81.

Nolfo-Clements, L. E. "Habitat Selection by Nutria in a Freshwater Louisiana Marsh." *Southeastern Naturalist* 11 (2012): 183–204.

———. "Nutria Survivorship, Movement Patterns, and Home Ranges." *Southeastern Naturalist* 8 (2009): 399–410.

———. "Seasonal Variations in Habitat Availability, Habitat Selection, and Movement Patterns of *Myocastor coypus* on a Subtropical Freshwater Floating Marsh." PhD diss., Tulane University, New Orleans, 2006.

Nolfo, L. E., and E. E. Hammond. "A Novel Method for Capturing and Implanting Radio-transmitters in Nutria." *Wildlife Society Bulletin* 34 (2006): 104–10.

Norgress, R. E. "The History of the Cypress Lumber Industry in Louisiana." *Louisiana Historical Quarterly* 30 (1947): 979–1059.

Norris, J. D. "A Campaign against Feral Coypus (*Myocastor coypus* Molina) in Great Britain." *Journal of Applied Ecology* 4 (1967): 191–99.

———. "The Control of Coypus (*Myocastor coypus* Molina) by Cage Trapping." *Journal of Applied Ecology* 4 (1967): 167–89.

Norton, T. E. *The Fur Trade in Colonial New York, 1686–1776.* Madison: University of Wisconsin Press, 1974.

Novak, M. "Traps and Trap Research." In *Wild Furbearer Management and Conservation in North America*, edited by M. Novak, J. A. Baker, M. E. Obbard, and B. Malloch, 941–69. Toronto: Ontario Trappers Association, Ontario Ministry of Natural Resources, 1987.

"Nutria Added to List of Fur Bearers; Severance Tax Increased on Mink and Alligators in New Act." *Louisiana Game, Fur and Fish* 4, no. 8 (July 1946): 8.

Nutria Management Team (NMT; Chesapeake Nutria Project). *Chesapeake Bay Eradication Project: Strategic Plan.* United States Fish and Wildlife Service, February 2012.

Nyman, J. A., and R. H. Chabreck. "Fire in Coastal Marshes: History and Recent Concerns." In *Proceedings 19th Tall Timbers Fire Ecology Conference—Fire in Wetlands: A Management Perspective*, edited by S. I. Cerulean and R. T. Engstrom, 135–41. Tallahassee, FL: Tall Timbers Research, 1995.

Nyman, J. A., R. H. Chabreck, and N. W. Kinler. "Some Effects of Herbivory and 30 Years of Weir Management on Emergent Vegetation in a Brackish Marsh." *Wetlands* 13 (1993): 165–75.

Oakes, D. S. "Isle of Spice." *Central Manufacturing District Magazine*, n.d., 1–12.

Obbard, M. E. "Fur Grading and Pelt Identification." In *Wild Furbearer Management and Conservation in North America*, edited by M. Novak, J. A. Baker, M. E. Obbard, and B. Malloch, 717–826. Toronto: Ontario Trappers Association, Ontario Ministry of Natural Resources, 1987.

Obbard, M. E., J. G. Jones, R. Newman, A. Booth, A. J. Satterthwaite, and G. Linscombe. "Furbearer Harvests in North America." In *Wild Furbearer Management and Conservation in North America*, edited by M. Novak, J. A. Baker, M. E. Obbard, and B. Malloch, 1007–34. Toronto: Ontario Trappers Association, Ontario Ministry of Natural Resources, 1987.

O'Neil, T. "From Eight Million Muskrats: The Fur Industry in Retrospect." *Louisiana Conservationist* 20 (1968): 9–14.

———. "The Fur Industry in Retrospect." *Louisiana Conservationist* 20 (1968): 9–14.

———. "Fur Luxury from Louisiana." *Louisiana Conservationist* 23 (1971): 24–27.

———. "Fur Price Discouraging." *Louisiana Conservationist* 19 (1967): 11–21.

———. "Fur Resources Trend Upward." *Louisiana Conservationist* 17 (1965): 9.

———. "Louisiana's Fur Industry Today." *Louisiana Conservationist* 15 (1963): 1–6.

———. *The Muskrat in the Louisiana Coastal Marshes: A Study of the Ecological, Geological, Biological, Tidal, and Climatic Factors Governing the Production and Management of the Muskrat Industry in Louisiana.* New Orleans: Louisiana Department of Wildlife and Fisheries, Fur and Game Division, 1949.

O'Neil, T., and G. Linscombe. "The Fur Animals, the Alligator, and the Fur Industry in Louisiana." *Louisiana Wildlife and Fisheries Commission, Wildlife Education Bulletin* 109 (1977): 1–68.

Osborne, M. A. "Acclimatizing the World: A History of the Paradigmatic Colonial Science." *Osiris* 15 (2000): 135–51.

Osgood, W. H. "The Mammals of Chile." *Field Museum of Natural History, Zoology Series* 30 (1943): 1–268.

Page, C. A., V. T. Harris, and J. Durand. "A Survey of Virus in Nutria." *Southwest Louisiana Journal* 1 (1957): 207–10.

Palmer, T. S. *The Dangers of Introducing Noxious Animals and Birds.* Cambridge: Department of Agriculture, 1893.

Panel on Advancing the Science of Climate Change, National Research Council. *Advancing the Science of Climate Change.* Washington, DC: National Academies Press, 2010.

Peek, J. M. *A Review of Wildlife Management.* New York: Prentice-Hall, 1986.

Peloquin, E. P. "Growth and Reproduction of the Feral Nutria *Myocastor coypus* (Molina) Near Corvallis, Oregon." Master's thesis, Oregon State University–Corvallis, 1969.

Penland, S., and K. E. Ramsey. "Relative Sea-Level Rise in Louisiana and the Gulf of Mexico: 1908–1988." *Journal of Coastal Research* 6 (1990): 323–42.

Penn, G. H. "The Life History of *Porocephalus crotali*, a Parasite of the Louisiana Muskrat." *Journal of Parasitology* 28 (1942): 277–83.

Phelps, N. *The Longest Struggle: Animal Advocacy from Pythagoras to PETA.* New York: Lantern Books, 2007.

Pietrzyk-Walknowski, J. "Sexual Maturation and Reproduction in the Nutria *Myocastor coypus*. III. The Testicle." *Folia Biologica (Warsaw)* 4 (1956): 151–62.

Poché, R. M. "Recent Studies with Norway Rats." *Proceedings Vertebrate Pest Conference* 18 (1998): 254–61.

———. "Status of Bromadiolone in the United States." *Proceedings Vertebrate Pest Conference* 12 (1986): 6–15.

"Police to Investigate Man Who Killed 4ft Rat." *Sky*, April 23, 2012, https://web.archive.org/ web/20120425231520/http://tyneandwear.sky.com/news/article/18835.

Prichard, W. "The Effects of the Civil War on the Louisiana Sugar Industry." *Journal of Southern History* 5 (1939): 315–32.

Pridham, T. J., J. Budd, and L. H. A. Karstad. "Common Diseases of Furbearing Mammals. II. Diseases of Chinchilla, Nutria, and Rabbits." *Canadian Veterinary Journal* 7 (1966): 84–87.

Quilter, J. *The Civilization of the Incas.* New York: Rosen, 2012.

Ramsey, C. W. "Nutria (*Myocastor coypu*) Fact Sheet." *Texas Agricultural Extension Service, Texas A&M University, College Station* (1975): L-1363.

———. "Rats to Riches." *Saturday Evening Post*, May 8, 1943, 14, 78–80.

"The Rat That Ate Louisiana." *Newsweek*, March 7, 1993, http://www.newsweek.com/ rat-ate-louisiana-190994.

Ray, A. J. *The Canadian Fur Trade in the Industrial Age.* Toronto: University of Toronto Press, 1990.

———. "The Fur Trade in North America: An Overview from a Historical Geographical Perspective." In *Wild Furbearer Management and Conservation in North America*, edited by M. Novak, J. A. Baker, M. E. Obbard, and B. Malloch, 21–30. Toronto: Ontario Trappers Association, Ontario Ministry of Natural Resources, 1987.

———. *Indians in the Fur Trade: Their Role as Trappers, Hunters, and Middlemen in the Lands Southwest of Hudson Bay, 1660–1870.* Toronto: University of Toronto Press, 1974.

Reggiani, G., L. Boitani, S. D'Antoni, and R. De Stefano. "Biology and Control of the Coypu in the Mediterranean Area." *Supplement Alle Ricerche di Biologia Della Selvaggina* 21 (1993): 67–100.

Reggiani, G., L. Boitani, and R. De Stefano. "Population Dynamics and Regulation in the Coypu *Myocastor coypus* in Central Italy." *Ecography (Copenhagen)* 18 (1995): 138–46.

Reyes, E., M. L. White, J. F. Martin, G. P. Kemp, J. W. Day, and V. Aravamuthan. "Landscape Modeling of Coastal Habitat Change in the Mississippi Delta." *Ecology* 81 (2000): 2331–49.

Rich, E. E. *The History of the Hudson Bay Company, 16701870, with a Foreword by Winston Churchill*. London: Hudson's Bay Record Society, 1958–59.

———. "Russia and the Colonial Fur Trade." *Economic History* 7 (1955): 3–8.

Rich, N. "The Most Ambitious Lawsuit Ever." *New York Times Magazine*, October 3, 2014, http://www.nytimes.com/interactive/2014/10/02/magazine/mag-oil-lawsuit.html?_r=0.

Roach, J. "Is Global Warming Making Hurricanes Worse?" *National Geographic News*, August 4, 2005.

Roberts, T. H., and D. H. Arner. "Food Habits of Beaver in East-Central Mississippi." *Journal of Wildlife Management* 48 (1984): 1414–19.

Robicheaux, B. L. "Ecological Implications of Variably Spaced Ditches on Nutria in a Brackish Marsh, Rockefeller Refuge, Louisiana." Master's thesis, Louisiana State University, Baton Rouge, 1978.

Robicheaux, B., and G. Linscombe. "Effectiveness of Live-Traps for Capturing Furbearers in a Louisiana Coastal Marsh." *Proceedings of the Southeast Association of Game and Fish Commissioners* 32 (1978): 208–12.

Rodgers, C., and R. B. Rankin. *New York: The World's Capital City, Its Development and Contributions to Progress*. New York: HarperCollins, 1948.

Ross, A. *The Red River Settlement*. Minneapolis: Ross and Haines, 1957.

Ross, D. A. *Introduction to Oceanography*. New York: HarperCollins College, 1995.

Ross, R. B., and M. D. Blum "Hurricane Audrey 1957." *Monthly Weather Review* 85 (1957): 221–27.

Rowlands, I. W., and R. B. Heap. "Histological Observations on the Ovary and Progesterone Levels in the Coypu (*Myocastor coypus*)." In *Comparative Biology of Reproduction in Mammals*, edited by I. W. Rowlands, 335–52. London: Symposia of the Zoological Society of London, 1966.

Rurik, H. "Shadeaux Sees Neaux Shadow: It's Official! Pierre, Our Nutria, Says It Will Be a Longer Spring." *Daily Iberian* (New Iberia, LA), February 2, 2012.

Ryszkowski, L. "The Space Organization of Nutria (*Myocastor coypus*) Populations." *Symposia of the Zoological Society of London* 18 (1966): 259–65.

Saadoum, A., M. C. Cabrera, and P. Castellucio. "Fatty Acids, Cholesterol and Protein Content of Nutria (*Myocastor coypus*) Meat from an Intensive Production System in Uruguay." *Meat Science* (2006): 778–84.

Sasser, C. E. "Vegetation Dynamics in Relation to Nutrients in Floating Marshes in Louisiana, USA." PhD thesis, University of Utrecht, The Netherlands, 1994.

Sasser, C. E., M. D. Dozier, J. G. Gosselink, and J. M. Hill. "Spatial and Temporal Changes in Louisiana's Barataria Basin Marshes, 1945–1980." *Environmental Management* 10 (1986): 671–80.

Sasser, C. E., J. G. Gosselink, G. O. Holm Jr., and J. M. Visser. "Freshwater Tidal Wetlands of the Mississippi River Delta." In *Tidal Freshwater Wetlands*, edited by A. Barendregt, A. Baldwin, P. Meire, and D. Whigham, 167–78. Leiden, the Netherlands: Backhuys, 2009.

Sasser, C. E., J. G. Gosselink, and G. P. Shaffer. "Distribution of Nitrogen and Phosphorus in a Louisiana Freshwater Floating Marsh." *Aquatic Botany* 41 (1991): 317–31.

Sasser, C. E., J. G. Gosselink, E. M. Swenson, C. M. Swarzenski, and N. C. Leibowitz. "Vegetation, Substrate, and Hydrology in Floating Marshes in the Mississippi River Delta Plain Wetlands, USA." *Vegetation* 122 (1996): 129–42.

Sasser, C. E., J. G. Gosselink, E. M. Swenson, and D. E. Evers. "Hydrology, Vegetation, and Substrate of Floating Marshes in Sediment-Rich Wetlands of the Mississippi River Delta Plain, Louisiana, USA." *Wetlands Ecology* 3 (1995): 171–87.

Sasser, C. E., G. O. Holm Jr., J. M. Visser and E. M. Swenson. "Thin-Mat Marsh Enhancement Demonstration Project TE-36." Final report, School of the Coast and Environment, Louisiana State University (2004): TE-36.

Sasser, C. E., J. M. Visser, D. E. Evers, and J. G. Gosselink. "The Role of Environmental Variables on Interannual Variation in Species Composition and Biomass in a Subtropical Minerotrophic Floating Marsh." *Canadian Journal of Botany* 73 (1995): 413–24.

Sasser, C. E., J. M. Visser, E. Mouton, J. Linscombe, and S. B. Hartley. "Vegetation Types in Coastal Louisiana in 2007." U.S. Geological Survey Open-File Report (2008): 1224.

Saul, J. "T-Boy the Nutria Predicts That Spring Is Right around the Corner." *WGNO News*, February 2, 2015.

Scheltema, G., and H. Westerhuijs, eds. *Exploring Historic Dutch New York*. New York: Museum of the City of New York/Dover, 2011.

Scheuring, W., and E. Bratkowski. "Hematological Values in Nutria." *Medycyna Weterynaryjna* 32 (1976): 239–41.

Scheuring, W., and J. A. Madej. "Sarcosporidiosis in Nutria." *Medycyna Weterynaryjna* 32 (1976): 437–38.

Schitoskey, F., Jr., J. Evans, and G. K. LaVoie. "Status and Control of Nutria in California." *Vertebrate Pest Conference* 5 (1972): 15–17.

Schleifstein, M. "Bounty Hunters Making Dent in Nutria Damage, State Wildlife Officials Say." *New Orleans Times-Picayune*, January 16, 2014, http://www.nola.com/environment/index.ssf/2014/01/bounty_hunters_making_dent_in.html.

———. "Vitter Backs Off Logging Proposal. It Would Open Harvest of Cypress in Wetlands." *New Orleans Times-Picayune*, October 18, 2005.

Schmidt, K. "Nutria Burrowing in Area Levee." *Houma (LA) Today*, April 21, 2010.

Schwartz, S. I. *The Mismapping of America*. Rochester, NY: University of Rochester Press, 2008.

Shaffer, G. P., C. E. Sasser, J. G. Gosselink, and M. Rejmanek. "Vegetation Dynamics in the Emerging Atchafalaya Delta, Louisiana, USA." *Journal of Ecology* 80 (1992): 677–87.

Shaw, J. H. *Introduction to Wildlife Management*. New York: McGraw-Hill, 1985.

Sheffels, T. R., and M. Sytsma. *Report on Nutria Management and Research in the Pacific Northwest*. Portland, OR: Report prepared for the Center for Lakes and Reservoirs, 2007.

Sheffels, T. R., M. D. Sytsma, J. Carter, and J. D. Taylor. "Efficacy of Plastic Mesh Tubes in Reducing Herbivory Damage by the Invasive Nutria *(Myocastor coypus)* in an Urban Restoration Site." *Northwest Science* 88 (2014): 269–79.

Sheng, Y. P., A. Lapetina, and G. Ma. "The Reduction of Storm Surge by Vegetation Canopies: Three-Dimensional Simulations." *Geophysical Research Letters* 39 (2012): L20601.

Shirley, M. G., R. H. Chabreck, and G. Linscombe. "Foods of Nutria in Fresh Marshes of Southeastern Louisiana." In *Worldwide Furbearer Conference Proceedings*, edited by J. A. Chapman and D. Pursley, 517–30. Frostburg, MD, 1981.

Silliman, B. R., and M. D. Bertness. "A Trophic Cascade Regulates Salt Marsh Primary Production." *Proceedings of the National Academy of Sciences of the USA* 99 (2002): 10500–10505.

Silliman, B. R., E. D. Grosholz, and M. D. Bertness, eds. *Human Impacts on Salt Marshes: A Global Perspective.* Berkeley: University of California Press, 2009.

Silver, T. *A New Face on the Countryside: Indians, Colonists, and Slaves in the South Atlantic Forests, 1500–1800.* Cambridge: Cambridge University Press, 1990.

Simpson, G. G. "The Principles of Classification and a Classification of Mammals." *Bulletin of the American Museum of Natural History* 85 (1945): 1–350.

Simpson, T. R., and W. G. Swank. "Trap Avoidance by Marked Nutria: A Problem in Population Estimation." *Proceedings, Annual Conferences, Southeast Association of Fish and Wildlife Agencies* 33 (1980): 11–14.

Singer, P. *Animal Liberation.* New York: Random House, 1975. Repr., New York: Ecco, 2002.

Six, J. "Biologist Says Nutria Have Arrived in New Jersey Marshland." NJ.com, December 5, 2007, http://www.nj.com/south/index.ssf/2007/12/biologist_says_nutria_have_arr.html.

Skowron-Cendrzak, A. "Sexual Maturation and Reproduction in *Myocastor coypus*. I. The Oestrus Cycle." *Folia Biologica (Warsaw)* 4 (1956): 119–38.

Smith, A. D. H. *John Jacob Astor: Landlord of New York.* New York: Cosimo, 2005.

Smith, D. S. "Foreign Birds for London Parks." *Avicultural Magazine* 5 (1906): 48–50.

Smith, L. "Louisiana Longleaf: An Endangered Legacy." *Louisiana Conservationist* 3 (1991): 24–27.

Smith, S. "Fur Trappers Are Taking On the Scourge of the Marshlands." *New York Times*, May 28, 2002.

Smith, W. R. *Brief History of the Louisiana Territory.* Saint Louis, MO: New, 1904.

Souther, R. F., and G. P. Shaffer. "The Effects of Submergence and Light on Two Age Classes of Baldcypress (*Taxodium distichum*) Seedlings." *Wetlands* 20 (2000): 697–706.

Southwick Associates. *Potential Economic Losses Associated with Uncontrolled Nutria Populations in Maryland's Portion of the Chesapeake Bay.* Prepared for the Maryland Department of Natural Resources (MDNR), November 2, 2004.

Spence, L. *Myths of the North American Indians.* New York: Gramercy Books, 1994.

Spiller, S. F., and R. H. Chabreck. "Wildlife Populations in Coastal Marshes Influenced by Weirs." *Proceeding of the Annual Conference Southeastern Association of Fish and Wildlife Commissioners* 29 (1975): 518–25.

St. Amant, L. S. *Louisiana Wildlife Inventory and Management Plan.* New Orleans: Pittman-Robertson Section, Fish and Game Division, Louisiana Wildlife and Fisheries Commission, 1959.

Stands In Timber, J., and M. Liberty. *Cheyenne Memories.* Lincoln: University of Nebraska, 1972.

Steed, B. C. "Why Don't We Just Shoot Them? An Institutional Analysis of Prairie Dog Protection in Iron County, Utah." Y673 Fall Mini-Conference (2005): 1–33.

Sterba, J. *Nature Wars: The Incredible Story of How Wildlife Comebacks Turned Backyards into Battlegrounds*. New York: Crown, 2012.

St. Johns Institute for Deaf-Mutes. *Our Young People*. Madison: State Historical Society of Wisconsin, 1904.

Stolzenburg, W. "Swan Song of the Ivory-Bill." *Nature Conservancy* 52 (2002): 38–47.

Stone, O. "Foreword." In *Monkey Business: The Disturbing Case That Launched the Animal Rights Movement*, by K. S.Guillermo, 9–12. New York: National Press Books, 1993.

Stubbe, M. "Die Nutria *Myocastor Coypu* in den Ostlichen Deutschen Bundeslandern." *Semi-aquatische Saugetiere* 9 (1992): 80–97.

———. "Die Nutria *Myocastor coypus* (Molina)." *Buch der Hege* 1 (1989): 630–39.

Sunder, J. E. *The Fur Trade on the Upper Missouri, 1840–1865*. Norman: University of Oklahoma Press, 1965.

Svihla, A., and R. D. Svihla. "The Louisiana Muskrat." *Journal of Mammalogy* 12 (1931): 12–28.

Tarver, J., G. Linscombe, and N. Kinler. *Fur Animals, Alligator, and the Fur Industry in Louisiana*. Baton Rouge: Miscellaneous Publications of the Fur and Refuge Division, Louisiana Department of Wildlife and Fisheries, 1987.

Taylor, D., and W. Neal. "Management Implications of Size-Class Frequency Distributions in Louisiana Alligator Populations." *Wildlife Society Bulletin* 12 (1984): 312–19.

Taylor, G. C. "The Saga of Petit Anse Island." *Attakapas Gazette* 19 (1984): 159–64.

Taylor, K. L., and J. B. Grace. "The Effects of Vertebrate Herbivory on Plant Community Structure in the Coastal Marshes of the Pearl River, Louisiana, USA." *Wetlands* 15 (1995): 68–73.

Taylor, K. L., J. B. Grace, G. R. Guntenspergen, and A. L. Foote. "The Interactive Effects of Herbivory and Fire on an Oligohaline Marsh, Little Lake, Louisiana, USA." *Wetlands* 14 (1994): 82–87.

Taylor, K. L., J. B. Grace, and B. D. Marx. "The Effects of Herbivory on Neighbor Interactions along a Coastal Marsh Gradient." *American Journal of Botany* 84 (1997): 709–15.

Taylor, Major G. R. *Deconstruction and Reconstruction*. New York: D. Appleton, 1879.

Taylor, M. B. *Canadian History: A Reader's Guide*. Vol. 1. *Beginnings to Confederation*. Toronto: University of Toronto Press, 1994.

Thomson, J. *The Works of James Thomson by James Thomson*. Vol. 2. 1763.

Thrapp, D. L. *Encyclopedia of Frontier Biography: A–F*. Lincoln: University of Nebraska Press, 1991.

Travis, H. G., and P. J. Schaible. "Effects of Diethylstilbestrol Fed Periodically during Gestation of Female Mink upon Reproductive and Kit Performance." *American Journal of Veterinary Research* 23 (1962): 359–61.

Tidwell, M. *The Ravaging Tide: Strange Weather, Future Katrinas, and the Coming Death of America's Coastal Cities*. New York: Free Press, 2006.

Todd, K. *Tinkering with Eden: A Natural History of Exotic Species in America*. New York: W. W. Norton, 2002.

Trillin, C. "The Nutria Problem." *Atlantic Monthly*, February 1995, 30–42.

Tucker, G. R. "La Salle Lands in Texas: La Salle and the Historians." *East Texas Historical Journal* 48 (2010): 40–58.

Turner, F. J. *The Significance of the Frontier in American History*. Edited by H. Simonson. New York: Frederick Ungar, 1963.

Valentine, J. M., Jr., J. R. Walther, K. M. McCartney, and L. M. Ivy. 1972. "Alligator Diets on the Sabine National Wildlife Refuge, Louisiana." *Journal of Wildlife Management* 36 (1972): 809–15.

Van Pelt, A. "Nutria—New Fur Bearer Increasing in Numbers in State." *Louisiana Conservationist* 4 (1946): 4–8.

Vaughn, G. *A Social History of the American Alligator: The Earth Trembles with His Thunder.* New York: St. Martin's Press, 1991.

Vavasseur, P. *Guide du Promeneur au Jardin Zoologique d'Acclimatation.* Paris: Jardin Zoologique d'Acclimatation, 1861.

Visser, J. M., C. E. Sasser, R. H. Chabreck, and R. G. Linscombe. "Long-Term Vegetation Change in Louisiana Tidal Marshes, 1968–1992." *Wetlands* 19 (1999): 168–75.

Visser, J. M., G. D. Steyer, G. P. Shaffer, S. S. Höppner, M. W. Hester, E. Reyes, P. Keddy, I. A. Mendelssohn, C. E. Sasser, and C. Swarzenski. "Louisiana Coastal Area (LCA) Ecosystem Restoration Study." In *Hydrodynamic and Ecological Modeling*. Vol. 4, app. C, "Habitat Switching Module," C143–C159. Baton Rouge: LCA, 2004.

Vogl, R. "Effects of Fire on the Plants and Animals of a Florida Wetland." *American Midland Naturalist* 89 (1973): 334–47.

Wade, D. A., and C. W. Ramsey. *Identifying and Managing Aquatic Rodents in Texas: Beaver, Nutria and Muskrats.* College Station: Texas Agriculture Extension Service, 1986.

Waldo, E. "Hurricane Damages Refuges." *Louisiana Conservationist* 9 (1957): 16–17.

———. "Storms and Wildlife." *Louisiana Conservationist* 13 (1961): 2–4.

Walker, J. R. "The Sun Dance and Other Ceremonies of the Oglala Division of the Teton Dakota." *Anthropological Papers of the American Museum of Natural History*. Vol. 16, pt. 2. New York: American Museum of Natural History, 1916.

Wallace, A. R. *Encyclopaedia Britannica*, 11th ed. Vol. 1, 114–21. 1911.

Walton, G. M., and H. Rockoff. *History of the American Economy.* Boston: Cengage Learning, 2009.

Wamsley, T. V., J. Atkinson, M. A. Cialone, A. S. Grzegorzewski, K. Dresback, R. Kolar, J. Westerink. "Influence of Wetland Degradation on Surge." *Proceedings of the 10th International Workshop on Wave Hindcasting and Forecasting and Coastal Hazard Symposium* (2007): 11–16.

Wang. F. C. "Effects of Levee Extension on Marsh Flooding." *Journal of Water Resources Planning Management* 113 (1987): 161–76.

Wanless, H. R. "Final Report and Findings from Technical Group, Envisioning the Future of the Gulf Coast Conference." New Orleans: America's Wetland: Campaign to Save Coastal Louisiana, 2006.

Warkentin, M. J. "Observations on the Behavior and Ecology of the Nutria in Louisiana." *Tulane Studies in Zoology and Botany* 15 (1968): 10–17.

Warwick, T. "Some Escapes of Coypus (*Myopotamus coypu*) from Nutria Farms in Great Britain." *Journal of Animal Ecology* 4 (1935): 146–47.

Washburn, M. "Evolution of the Trapping Industry." *Louisiana Conservationist* 4 (1951): 8–24.

———. "Plan Fur Industry Comeback." *Louisiana Conservationist* 5, no. 3 (November–December 1952): 12–25.

———. "The Trapper Calls It a Bad Day." *Louisiana Conservationist* 3, no. 6 (February 1951): 12–13.

Webb, W. P. *The Great Frontier.* Austin: University of Texas Press, 1964.

———. *The Great Plains.* New York: Grosset and Dunlap, 1931.

———. *The Texas Rangers.* Austin: University of Texas Press, 1935.

Weber, D. J. *The Taos Trapper: The Fur Trade in the Far Southwest, 1540–1846.* Norman: University of Oklahoma Press, 1970.

Webert, F. J. "Hurricane Damage to Fur Industry." *Louisiana Conservationist* 8 (1956): 10–13, 20–21.

Weir, B. J. "Reproductive Characteristics of Hystricomorph Rodents." *Symposia of the Zoological Society of London* 34 (1974): 265–301.

Wentz, W. A. "The Impact of Nutria (*Myocastor coypus*) on Marsh Vegetation in the Willamette Valley, Oregon." Master's thesis, Oregon State University–Corvallis, 1971.

White, P. S. "Synthesis: Vegetation Pattern And Process In The Everglades Ecosystem." In *Everglades: The Ecosystem and Its Restoration,* edited by S. Davis and J. Ogden, 445–446. Del Ray Beach, FL: St. Lucia Press, 1994.

White, R. *It's Your Misfortune and None of My Own: A New History of the American West.* Norman: University of Oklahoma Press, 1991.

———. *The Middle Ground: Indians, Empires, and Republics in the Great Lakes Region, 1650–1815.* Cambridge: Cambridge University Press, 1991.

White, S. "The Beaver as National Symbol: Why Is a Furry Mammal Still an Emblem of Canada?" *Huffington Post,* July 1, 2011, http://www.huffingtonpost.ca/2011/07/01/canadian-symbols-beaver_n_886777.html.

Wiebe, J., and E. Mouton. "Nutria Harvest and Distribution 2007–2008 and a Survey of Nutria Herbivory Damage in Coastal Louisiana in 2008." Fur and Refuge Division Louisiana Department of Wildlife and Fisheries (2008): 1–39.

Wilkinson, K. "'Try It—You'll Like It,' Chefs Says about Those Pesky Nutria." *New Orleans City Business* 19 (1999): 25–28.

Williams, B. K. "Logic and Science in Wildlife Biology." *Journal of Wildlife Management* 61 (1997): 1007–15.

Willner, G. R., J. A. Chapman, and D. Pursley. "Reproduction, Physiological Responses, Food Habits, and Abundance of Nutria on Maryland Marshes." *Wildlife Monographs* 65 (1979): 1–43.

Willner, G. R., K. R. Dixon, and J. A. Chapman. "Age Determination and Mortality of the Nutria (*Myocastor coypus*) in Maryland, USA." *Zeitschrift für Säugetierkunde* 48 (1983): 19–34.

Willner, G. R., K. R. Dixon, J. A. Chapman, and J. R. Stauffer Jr. "A Model for Predicting Age-Specific Body Weights of Nutria without Age Determination." *Journal of Applied Ecology* 7 (1980): 343–47.

———. "Nutria: *Myocastor coypus*." In *Wild Animals of North America,* edited by J. A. Chapman and G. A. Feldhamer, 1059–76. Baltimore: Johns Hopkins University Press, 1982.

Willner, G. R. G. A. Feldhamer, E. E. Zucker, and J. A. Chapman. "*Ondatra zibethicus.*" *American Society of Mammalogists, Mammalian Species* 141 (1980): 1–8.

Wilson, E. D., and A. A. Dewees. "Body Weights, Adrenal Weights and Oestrous Cycles of Nutria." *Journal of Mammalogy* 43 (1962): 362–64.

Wilson, E. D., M. X. Zarrow, and H. S. Lipscomb. "Bilateral Dimorphism of the Adrenal Glands in the Coypu (*Myocastor coypus*, Molina)." *Endocrinology* 74 (1964): 515–17.

Wishart, D. J. *The Fur Trade of the American West, 1807–1840*. Lincoln: University of Nebraska Press, 1992.

Wissler, C. "Societies of the Plains Indians." *Anthropological Papers of the American Museum of Natural History*. Vol. 11. New York: American Museum of Natural History, 1916.

Wolfe, J. L., D. K. Bradshaw, and R. H. Chabreck. "Alligator Feeding Habits: New Data and a Review." *Northeast Gulf Science* 9 (1987): 1–8.

Wolfe, M. L., and J. A. Chapman. "Principles of Furbearer Management." In *Wild Furbearer Management and Conservation in North America*, edited by M. Novak, J. A. Baker, M. E. Obbard, and B. Malloch, 101–12. Toronto: Ontario Trappers Association, Ontario Ministry of Natural Resources, 1987.

Woods, C. A. "Comparative Myology of Jaw, Hyoid, and Pectoral Appendicular Regions of New and Old World Hystricomorph Rodents." *Bulletin of the American Museum of Natural History* 147 (1972): 115–98.

———. "The History and Classification of South American Rodents: Reflections on the Far Away and Long Ago." In *Mammalian Biology in South America*, edited by M. A. Mares and H. H. Genoways, 377–92. Pittsburgh, PA: Special Publications Series, Pymatuning Laboratory of Ecology, University of Pittsburgh, 1982.

———. "How Hystricognath Rodents Chew." *American Zoologist* 16 (1976): 215.

———. "Hystricognath Rodents." In *Orders and Families of Recent Mammals of the World*, edited by S. Anderson and J. K. Jones Jr., 389–446. New York: John Wiley and Sons, 1984.

Woods, C. A., L. Contreras, G. Willner-Chapman, and H. P. Whidden. "*Myocastor coypus.*" *American Society of Mammalogists, Mammalian Species* 398 (1992): 1–8.

Woods, C. A., and E. B. Howland. "Adaptive Radiation of Capromyid Rodents: Anatomy of the Masticatory Apparatus." *Journal of Mammalogy* 60 (1979): 95–116.

———. "The Skin Musculature of Hystricognath and Other Selected Rodents." *Zentralblatt fur Veterinaermedizin Reihe C* 6 (1977): 240–64.

Woodward, C. V. *Origins of the New South*. Baton Rouge: Louisiana State University Press, 1971.

Woodward, S. L., and J. A. Quinn. *Encyclopedia of Endangered Species: From Africanized Honey Bees to Zebra Mussels*. Santa Barbara, CA: ABC-CLIO, 2011.

Woolfolk, D., ed. *Tangipahoa Crossings: Excursions into Tangipahoa History*. Baton Rouge, LA: Moran, 1979.

Wright, J. V. "Archaeological Evidence for the Use of Furbearers in North America." In *Wild Furbearer Management and Conservation in North America*, edited by M. Novak, J. A. Baker, M. E. Obbard, and B. Malloch, 3–12. Toronto: Ontario Trappers Association, Ontario Ministry of Natural Resources, 1987.

# Index